W9-CRZ-137

ELECTRON-PHONON INTERACTIONS IN NOVEL NANOELECTRONICS

ELECTRON-PHONON INTERACTIONS IN NOVEL NANOELECTRONICS

TAKASHI KATO

Nova Science Publishers, Inc.
New York

For permission to use material from this book please contact us:
Telephone 631-231-7269; Fax 631-231-8175
Web Site: http://www.novapublishers.com

NOTICE TO THE READER
The Publisher has taken reasonable care in the preparation of this book, but makes no expressed or implied warranty of any kind and assumes no responsibility for any errors or omissions. No liability is assumed for incidental or consequential damages in connection with or arising out of information contained in this book. The Publisher shall not be liable for any special, consequential, or exemplary damages resulting, in whole or in part, from the readers' use of, or reliance upon, this material.

Independent verification should be sought for any data, advice or recommendations contained in this book. In addition, no responsibility is assumed by the publisher for any injury and/or damage to persons or property arising from any methods, products, instructions, ideas or otherwise contained in this publication.

This publication is designed to provide accurate and authoritative information with regard to the subject matter covered herein. It is sold with the clear understanding that the Publisher is not engaged in rendering legal or any other professional services. If legal or any other expert assistance is required, the services of a competent person should be sought. FROM A DECLARATION OF PARTICIPANTS JOINTLY ADOPTED BY A COMMITTEE OF THE AMERICAN BAR ASSOCIATION AND A COMMITTEE OF PUBLISHERS.

Library of Congress Cataloging-in-Publication Data

ISBN: 978-1-60692-170-8

Available upon request

Published by Nova Science Publishers, Inc. ⌂ *New York*

CONTENTS

Preface		vii
Chapter I	Introduction	1
Chapter II	Theoretical Background	5
Chapter III	Electron–Phonon Coupling Constants for the Charged Electronic States of Polyacenes, Polyfluoroacenes, B, N-Substituted Polyacenes, and Polycyanodienes	15
Chapter IV	Optimized Structures	21
Chapter V	Electron–Phonon Coupling Constants	27
Chapter VI	Total Electron–Phonon Coupling Constants	49
Chapter VII	The Logarithmically Averaged Phonon Frequencies	63
Chapter VIII	Concluding Remarks	73
Acknowledgment		77
References		79
Index		85

PREFACE

In this book, the electron–phonon interactions in the charged molecular systems such as polyacenes, polyfluoroacenes, B, N-substituted polyacenes, and polycyanodienes are discussed. We estimated the electron–phonon coupling constants and the frequencies of the vibronic active modes playing an essential role in the electron–phonon interactions in order to discuss how CH–CF, CC–BN, and CC–CN substitutions are closely related to the essential characteristics of the electron–phonon interactions in these molecules by comparing the calculated results for charged polyacenes with those for charged B, N-substituted polyacenes and polycyanodienes, respectively. The C–C stretching modes around 1500 cm^{-1} strongly couple to the highest occupied molecular orbitals (HOMO), and the lowest frequency modes and the C–C stretching modes around 1500 cm^{-1} strongly couple to the lowest unoccupied molecular orbitals (LUMO) in polyacenes. The C–C stretching modes around 1500 cm $^{-1}$ strongly couple to the HOMO and LUMO in polyfluoroacenes. The B–N stretching modes around 1500 cm $^{-1}$ strongly couple to the HOMO and LUMO in B, N-substituted polyacenes. The C–C and C–N stretching modes around 1500 cm^{-1} strongly couple to the HOMO and LUMO in polycyanodienes. The total electron–phonon coupling constants for the monocations (l_{HOMO}) and monoanions (l_{LUMO}) decrease with an increase in molecular size in polyacenes, polyfluoroacenes, B, N-substituted polyacenes, and polycyanodienes. In general, we can expect that monocations and monoanions, in which number of carriers per atom is larger, affords larger value. The CH–CF, CC–BN, and CC–CN atomic substitutions are effective way to seek for larger l_{HOMO} values, and the CH–CF and CC–CN atomic substitutions are effective way to seek for larger l_{LUMO} values in polyacenes. The logarithmically averaged phonon frequencies ($\omega_{\ln}$) which measure the frequencies of the vibronic active modes playing an essential role in the electron–phonon interactions for the monocations $(\omega_{\mathrm{ln,HOMO}})$ and monoanions $(\omega_{\mathrm{ln,LUMO}})$ in polyacenes, polyfluoroacenes, B, N-substituted polyacenes, and polycyanodienes are investigated. The $\omega_{\mathrm{ln,HOMO}}$ values decrease with an increase in molecular size in polyacenes, polyfluoroacenes, and polycyanodienes, and the $\omega_{\mathrm{ln,LUMO}}$ values decrease with an increase in molecular size in polyacenes, polyfluoroacenes, B, N-substituted polyacenes, and polycyanodienes. We can expect that in the hydrocarbon molecular systems, the ω_{ln} values would basically decrease by substituting hydrogen atoms by heavier atoms. This can be understood from the fact that the frequencies of all vibronic active modes in polyacenes downshift by H–F substitution. However, considering that the ω_{ln} value for the LUMO rather localized on carbon atoms in large sized polyfluoroacenes becomes larger by H–F substitution, we can expect that the ω_{ln} value for a molecular orbital localized on carbon atoms has a possibility to increase by substituting hydrogen atoms by heavier atoms if the phase patterns of the molecular orbital do not significantly change by such atomic substitution. Therefore, the detailed properties of the vibrational modes and the electronic structures as well as the molecular weights are closely related to the frequencies of the vibronic active modes playing an important role in the electron–phonon interactions in the monoanions of polyfluoroacenes.

Chapter I

INTRODUCTION

In modern physics and chemistry, the effects of vibronic interactions [1] and electron–phonon interactions [1–3] in molecules and crystals have been an important topic. Analysis of vibronic interaction [1–3] is important for the prediction of electronic control of nuclear motions in degenerate electronic systems. Application of vibronic interaction theory covers a large variety of research fields such as spectroscopy,[4] instability of molecular structure, electrical conductivity,[5] and superconductivity.[5, 6] Vibronic interactions in discrete molecules can be viewed as the coupling between frontier orbitals and molecular vibrations, while those in solids are the coupling between free electrons near the Fermi level and acoustic phonons. There is a close analogy between them.

Electron–phonon coupling [1–3] is the consensus mechanism for attractive electron–electron interactions in the Bardeen–Cooper–Schrieffer (BCS) theory of superconductivity.[5,6] Since Little's proposal for a possible molecular superconductor based on exciton mechanism,[7] the superconductivity of molecular systems has been extensively investigated. Although such a unique mechanism has not yet been established, advances in design and synthesis of molecular systems have yielded a lot of BEDT-TTF-type organic superconductors, [8, 9] where BEDT-TTF is *bis*(ethylenedithio)tetrathiafulvalene. An inverse isotope effect due to substituting hydrogen by deuterium in organic superconductivity was observed by Saito et al.[10] Goddard et al. proposed that the mechanism for superconductivity of BEDT–TTF type organic molecules involves the coupling of charge transfer to the boat deformation mode.[11] It was found that the alkali-doped A_3C_{60} complexes [12] exhibit superconducting transition temperatures (T_cs) of more than 30 K (Ref. [13]) and 40 K under pressure.[14] In superconductivity in alkali-doped fullerenes,[15] pure intramolecular Raman-active modes have been suggested to be important in a BCS-type [6] strong coupling scenario.

It was proposed that the electron–phonon interactions dominate the charge transport in the crystals of naphthalene ($C_{10}H_8$) (2a), anthracene (3a), tetracene (4a), and pentacene (5a).[16–18] Interestingly, from a theoretical viewpoint, possible superconductivity of polyacene has been proposed.[19,20] Saito et al.[10] observed an inverse isotope effect due to substituting hydrogen by deuterium in organic superconductivity. It is important to consider how intramolecular or intermolecular vibrations play a role in the occurrence of superconductivity. If the intermolecular vibrations are important, the phonon-frequency dependence of the transition temperature appears in the prefactor through Debye frequency in the formula for the superconducting transition temperatures (T_cs) in the Bardeen–Cooper–Schrieffer (BCS) theory so that the normal isotope effects are expected. If the intramolecular vibrations are important, the phonon-frequency dependence appears in the denominator of the expression for the electron–phonon coupling constant so that the inverse isotope effect can be expected in some cases. From the inverse isotope effect on deuterium substitution observed by Saito et al.,[10] we expect that such inverse isotope effects can be widely observed in molecular organic superconductors. The origin of such inverse isotope effects in organic superconductors has not yet been fully elucidated. Shortly after the discovery of superconductivity in palladium hydrides,[21] an isotope effect in T_c was found by Stritzker and Buckel.[22] The T_c of Pd-D was higher than that of Pd-H,[23] contrary to the expectations from a simple BCS theory After that, an even larger inverse isotope effect for Pd-T was measured by Schirber et al.[24]

In previous work, we have analyzed the vibronic interactions and estimated possible T_cs in the monocations of polyacenes based on the hypothesis that the vibronic interactions between the intramolecular vibrations and the highest occupied molecular orbitals (HOMO) play an essential role in the occurrence of superconductivity in positively charged nanosized molecular systems.[25] On the basis of an experimental study of ionization spectra using the high-resolution gas-phase photoelectron spectroscopy, the electron–phonon interactions in the positively charged polyacenes were well studied recently.[26] Our predicted frequencies for the vibrational modes which play an essential role in the electron–phonon interactions [25] as well as the predicted total electron–phonon coupling constants [25] are in excellent agreement with those obtained from the experimental research.[26]

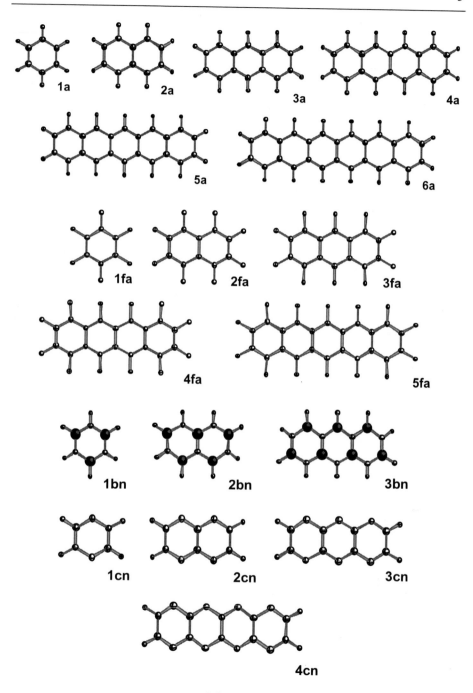

Scheme 1.

In this book, we discuss the electron–phonon interactions in the charged molecular systems such as polyacenes such as C_6H_6 (1a), $C_{10}H_8$ (2a), $C_{14}H_{10}$ (3a), $C_{18}H_{12}$ (4a), and $C_{22}H_{14}$ (5a),[25] polyfluoroacenes such as C_6F_6 (1fa), $C_{10}F_8$ (2fa), $C_{14}F_{10}$ (3fa), $C_{18}F_{12}$ (4fa), and $C_{22}F_{14}$ (5fa),[27] B, N-substituted polyacenes such as $B_3N_3H_6$ (1bn), $B_5N_5H_8$ (2bn), $B_7N_7H_{10}$ (3bn),[28] and polycyanodienes such as $C_4N_2H_6$ (1cn), $C_6N_4H_8$ (2cn), $C_8N_6H_{10}$ (3cn), and $C_{10}N_8H_{12}$ (4cn)[29] (Scheme 1). We will estimate the electron–phonon coupling constants and the frequencies of the vibronic active modes playing an essential role in the electron–phonon interactions. These physical values are essential to discuss the several physical phenomena such as intramolecular electrical conductivity, intermolecular charge transfer, attractive electron–electron interactions and Bose–Einstein condensation, and superconductivity, which will be discussed in detail in the next review article.

Motivated by the possible inverse isotope effects in Pd-H, Pd-D, and Pd-T superconductivity [21–24] and organic superconductivity observed by Saito et al.,[10] we discuss how the H–F substitution are closely related to the essential characteristics of the electron–phonon interactions in these molecules by comparing the calculated results for charged polyacenes with those for charged polyfluoroacenes, since fluorine atoms are much heavier than D and T atoms, and the phase patterns of the frontier orbitals such as the HOMO and LUMO are not expected to be significantly changed. Furthermore, we discuss how C–BN and C–N substitutions are closely related to the essential characteristics of the electron–phonon interactions in these molecules by comparing the calculated results for charged polyacenes with those for charged B, N-substituted polyacenes and polycyanodienes, respectively. We can expect that the characteristics of electron–phonon interactions are significantly changed by such atomic substitutions because of electronegativity perturbation[30] in polyacenes. These physical values are essential to discuss the several physical phenomena such as intramolecular electrical conductivity, intermolecular charge transfer, attractive electron–electron interactions and Bose–Einstein condensation, and superconductivity, which will be discussed in detail in the next review article.

Chapter II

THEORETICAL BACKGROUND

We describe the theoretical background for the vibronic coupling in polyacenes, polyfluoroacenes, B, N-substituted polyacenes, and polycyanodienes. We will use small letters for "one-electron orbital symmetries" and capital letters for symmetries of both "electronic" and "vibrational" states, as usual. The vibronic matrix element, $E_{xy}(r,Q)$, [1–3] is given by

$$E_{xy}(r,Q) = \varepsilon_{xy}(r,Q) - \varepsilon_{xy}(r,0) \overset{=}{\ } \sum_{\alpha}\left(\frac{\partial \varepsilon_{xy}}{\partial Q_\alpha}\right)_0 Q_\alpha + \frac{1}{2}\sum_{\alpha,\beta}\left(\frac{\partial^2 \varepsilon_{xy}}{\partial Q_\alpha \partial Q_\beta}\right)_0 Q_\alpha Q_\beta \tag{1}$$

where $\varepsilon_{xy}(r,Q)$ is defined as

$$\varepsilon_{xy}(r,Q) = \left\langle \phi_x \middle| h(r,Q) \middle| \phi_y \right\rangle \tag{2}$$

Here, $h(r,Q)$ is the Hamiltonian of one-electron orbital energy, and ϕ_x and ϕ_y are one-electron wave functions. r and Q signify the whole set of coordinates of the electrons and nuclei, respectively. What we see in the first term on the right-hand side of Eq. (1) is the linear orbital vibronic coupling constant.

We discuss a theoretical background for the orbital vibronic interactions in polyacenes, polyfluoroacenes, B, N-substituted polyacenes, and polycyanodienes. The potential energy for the neutral ground state, negatively and positively charged electronic states in polyacenes, polyfluoroacenes, B, N-substituted polyacenes, and polycyanodienes are shown in Figure 1. Here, we take a one-electron approximation into account; the vibronic coupling constants of the vibrational modes to the electronic states in the monoanions and cations of polyacenes, polyfluoroacenes, B, N-substituted polyacenes, and polycyanodienes are defined as a sum of orbital vibronic coupling constants from all the occupied orbitals, [1(a)]

$$g_{\text{electronic state}} = \sum_i^{\text{occupied}} g_i \qquad (3)$$

(a) Ground states in neutral polyacenes, polyfluoroacenes, B, N-substituted polyacenes, and polycyanodienes

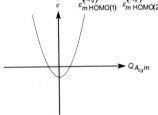

(b) The monocations of **1a**, **1fa**, and **1bn**

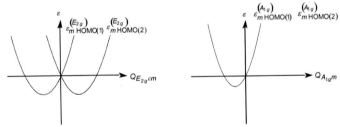

(c) The monocations of D_h symmetric polyacenes, polyfluoroacenes, and polycyanodienes, ar C_{2v} symmetric B, N-substituted polyacenes

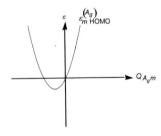

Figure 1. Potential energy for the neutral ground state and positively charged electronic states in polyacenes, polyfluoroacenes, B, N-substituted polyacenes, and polycyanodienes.

Considering the one-electron approximation and that the first derivatives of the total energy vanish in the ground state at the equilibrium structure in neutral polyacenes, polyfluoroacenes, B, N-substituted polyacenes, and polycyanodienes (i.e., $g_{\text{neutral}} = \sum_i^{\text{HOMO}} g_i = 0$) (Figure 1 (a)), and one electron must be injected into (removed from) the LUMO (HOMO) to generate the monoanions (monocations), the vibronic coupling constants of the totally symmetric vibrational modes to the electronic states of the monoanions and cations of polyacenes, polyfluoroacenes, B, N-substituted polyacenes, and polycyanodienes can be defined by Eqs. (4) and (5), respectively,

$$g_{\text{monoanion}}(\omega_m) = g_{\text{LUMO}}(\omega_m), \tag{4}$$

$$g_{\text{monocation}}(\omega_m) = g_{\text{HOMO}}(\omega_m). \tag{5}$$

A. Vibronic Interactions between the Twofold Degenerate Frontier Orbitals and the E_{2g} Vibrational Modes In Benzene

Benzene (**1a**) has the twofold degenerate HOMO and LUMO due to its high D_{6h} symmetry. The symmetry labels of the HOMO and LUMO are e_{1g} and e_{2u}, respectively. Now we take two approximations into account in Eq. (1). First, we ignore the nondiagonal matrix elements containing the quadratic vibronic constants; only the terms of type Q_α^2 are taken into account. Second, we focus upon the diagonal processes; we consider the direct product of the orbital symmetries, which can be reduced as

$$e_{1g} \times e_{1g} = e_{2u} \times e_{2u} = A_{1g} + A_{2g} + E_{2g}. \tag{6}$$

Thus, the A_{1g} and E_{2g} modes can linearly couple to the e_{1g} HOMO and e_{2u} LUMO. The numbers of the A_{1g} and E_{2g} modes are 2 and 4, respectively. We must consider multimode problems, but in the limit of linear vibronic coupling we can treat each twofold degenerate set of modes (mode index m as shown below) independently.[1]

Let us first consider the orbital vibronic coupling of the twofold degenerate e_{1g} HOMO to the E_{2g} vibrational mode. Taking two approximations mentioned above and Eq. (1) into account, the two energy sheets of the e_{1g} HOMO that can couple to the mth E_{2g} vibrational mode are given in the form of Eq. (7), where $Q_{E_{2g}\gamma m}$ is the vibrational normal coordinate belonging to row γ of the vibrational mode of the irreducible representation E_{2g}:

$$\varepsilon_{mxy}^{(E_{2g})}\left(Q_{E_{2g}\theta m}, Q_{E_{2g}\varepsilon m}\right)$$
$$= \frac{1}{2}\sum_{\gamma} K_{E_{2g}m} Q_{E_{2g}\gamma m}^2 \delta_{xy} + \sum_{\gamma} \left\langle e_{1gx}\,\text{HOMO}\left|\left(\frac{\partial h}{\partial Q_{E_{2g}\gamma m}}\right)_0\right| e_{1gy}\,\text{HOMO}\right\rangle Q_{E_{2g}\gamma m}$$
$$(m = 1,2,...,4) \tag{7}$$

Here, γ takes two vibrational states θ and ε. The first term on the right-hand side of Eq. (7) is the elastic term, and $\left\langle e_{1gx}\,\text{HOMO}\left|\left(\frac{\partial h}{\partial Q_{E_{2g}\gamma m}}\right)_0\right| e_{1gy}\,\text{HOMO}\right\rangle$ in the

second term corresponds to the linear orbital vibronic coupling constant, which is a good measure of the interaction between the vibration labeled by γ in the mth mode of the irreducible representation E_{2g} and the molecular orbitals labeled by x and y in the irreducible representation e_{1g}. According to the Wigner–Eckart theorem, [1] we can rewrite the second term on the right-hand side of Eq. (7) using the Clebsch–Gordan coefficients and we obtain the vibronic coupling matrix[1] as

$$\varepsilon_{E_{2g}m} = A_m \begin{pmatrix} -Q_{E_{2g}\varepsilon m} & Q_{E_{2g}\theta m} \\ Q_{E_{2g}\theta m} & Q_{E_{2g}\varepsilon m} \end{pmatrix}, \tag{8}$$

where A_m corresponds to the reduced matrix element for the mth mode of vibration:

$$A_m = \left\langle e_{1g}\,\text{HOMO}\left\|\left(\frac{\partial h}{\partial Q_{E_{2g}m}}\right)_0\right\| e_{1g}\,\text{HOMO}\right\rangle \tag{9}$$

A_m depends on the irreducible representation and not on rows. Using Eq. (8), Eq. (7) can be transformed into Eq. (10):

$$\varepsilon_m^{(E_{2g})}\left(Q_{E_{2g}\theta m}, Q_{E_{2g}\varepsilon m}\right) = \frac{1}{2}K_{E_{2g}m}\left(Q_{E_{2g}\theta m}^2 + Q_{E_{2g}\varepsilon n}^2\right)\begin{pmatrix} 1 & 0 \\ 0 & 1 \end{pmatrix} + A_m\begin{pmatrix} -Q_{E_{2g}\varepsilon m} & Q_{E_{2g}\theta m} \\ Q_{E_{2g}\theta m} & Q_{E_{2g}\varepsilon m} \end{pmatrix}$$

(10)

Now we consider one member of the twofold degenerate vibrational mode. For example, when we consider only the $Q_{E_{2g}\varepsilon n}$ mode ($Q_{E_{2g}\theta m} = 0$), the energy sheets of the molecular orbital in row ε of the mth mode become

$$\varepsilon_m^{(E_{2g})}\left(Q_{E_{2g}\varepsilon n}\right) = \frac{1}{2}K_{E_{2g}m}Q_{E_{2g}\varepsilon m}^2\begin{pmatrix} 1 & 0 \\ 0 & 1 \end{pmatrix} + A_m\begin{pmatrix} -Q_{E_{2g}\varepsilon m} & 0 \\ 0 & Q_{E_{2g}\varepsilon m} \end{pmatrix}$$

(11)

This has already been diagonalized, and the energy sheets of the molecular orbital (HOMO (1) and HOMO (2)) in row ε cut by the plane $Q_{E_{2g}\theta m} = 0$ become

$$\varepsilon_{m\,\text{HOMO (1)}}^{(E_{2g})}\left(Q_{E_{2g}\varepsilon m}\right) = \frac{1}{2}K_{E_{2g}m}Q_{E_{2g}\varepsilon m}^2 + A_mQ_{E_{2g}\varepsilon m}$$

(12)

and

$$\varepsilon_{m\,\text{HOMO (2)}}^{(E_{2g})}\left(Q_{E_{2g}\varepsilon n}\right) = \frac{1}{2}K_{E_{2g}m}Q_{E_{2g}\varepsilon m}^2 - A_mQ_{E_{2g}\varepsilon n}$$

(13)

$$Q_{E_{2g}\theta m} = 0$$

(14)

These are illustrated in Figure 1. Only the cases where the quadratic vibronic constants are positive are shown as examples here. We see from this illustration that the E_{2g} modes lift the degeneracy of the twofold degenerate orbitals. The dimensionless diagonal linear orbital vibronic coupling constants of the e_{1g} HOMO for its mth mode is defined by Eq. (15).

$$g_{e_{1g}\,\text{HOMO}}(\omega_m) = \frac{1}{\tilde{\omega}_m}\left\langle e_{1g}\,\text{HOMO}\left\|\left(\frac{\partial h}{\partial q_{E_{2g}m}}\right)_0\right\| e_{1g}\,\text{HOMO}\right\rangle$$

(15)

In these equations, $q_{E_{2g}m}$ is the dimensionless normal coordinate [31] defined by

$$q_{E_{2g}m} = \sqrt{\omega_m / \hbar} Q_{E_{2g}m} \qquad (16)$$

Similar discussions can be made in the vibronic interactions between the E_{2g} modes and the e_{2u} LUMO, and the dimensionless diagonal linear orbital vibronic coupling constants of the e_{2u} LUMO for the mth mode is defined by

$$g_{e_{2u}\text{ LUMO}}(\omega_m) = \frac{1}{\hbar\omega_m}\left\langle e_{2u}\text{ LUMO}\left\|\left(\frac{\partial h}{\partial q_{E_{2g}m}}\right)_0\right\| e_{2u}\text{ LUMO}\right\rangle \qquad (17)$$

B. Vibronic Interactions between the Twofold Degenerate Frontier Orbitals and the A_{1g} Vibrational Modes in Benzene

In this case, the two energy sheets of the molecular orbitals are given in the form:

$$\varepsilon_m^{(A_{1g})}(Q_{A_{1g}m}) = \frac{1}{2}K_{A_{1g}m}Q_{A_{1g}m}^2\begin{pmatrix} 1 & 0 \\ 0 & 1 \end{pmatrix} + B_m\begin{pmatrix} Q_{A_{1g}m} & 0 \\ 0 & Q_{A_{1g}m} \end{pmatrix} \qquad (18)$$

where B_m corresponds to the reduced matrix element written as

$$B_m = \left\langle e_{1g}\text{ HOMO}\left\|\left(\frac{\partial h}{\partial Q_{A_{1g}m}}\right)_0\right\| e_{1g}\text{ HOMO}\right\rangle \qquad (19)$$

This has already been diagonalized, and the energy sheets of the molecular orbital (HOMO (1) and HOMO (2)) become

$$\varepsilon_m^{(A_{1g})}{}_{\text{HOMO (1)}}(Q_{A_{1g}m}) = \varepsilon_m^{(A_{1g})}{}_{\text{HOMO (2)}}(Q_{A_{1g}m}) = \frac{1}{2}K_{A_{1g}m}Q_{A_{1g}m}^2 + B_m Q_{A_{1g}m} \qquad (20)$$

These are illustrated in Figure 1. We can see from this illustration that the A_{1g} modes do not lift the degeneracy of the twofold degenerate orbitals. The dimensionless diagonal linear orbital vibronic coupling constants of the e_{1g} HOMO for its mth mode is defined by Eq. (21),

$$g_{e_{1g} \text{ HOMO}}(\omega_m) = \frac{1}{\hbar\omega_m} \left\langle e_{1g} \text{ HOMO} \left\| \left(\frac{\partial h}{\partial q_{A_{1g}m}} \right)_0 \right\| e_{1g} \text{ HOMO} \right\rangle \tag{21}$$

In this equation, $q_{A_{1g}m}$ is the dimensionless normal coordinate [31] of the mth mode defined by using the normal coordinate $Q_{A_{1g}m}$ as

$$q_{A_{1g}m} = \sqrt{\omega_m / \hbar} Q_{A_{1g}m} \tag{22}$$

In a similar way, the dimensionless diagonal linear orbital vibronic coupling constants of the e_{2u} LUMO for the mth mode is defined by Eq. (23),

$$g_{e_{2u} \text{ LUMO}}(\omega_m) = \frac{1}{\hbar\omega_m} \left\langle e_{2u} \text{ LUMO} \left\| \left(\frac{\partial h}{\partial q_{A_{1g}m}} \right)_0 \right\| e_{2u} \text{ LUMO} \right\rangle \tag{23}$$

C. Vibronic Interactions between the Nondegenerate Frontier Orbitals and the A_g Vibrational Modes in Polyacenes

Let us look into orbital vibronic coupling in polyacenes that have D_{2h} geometries, and the nondegenerate HOMOs and LUMOs, the symmetry labels of the HOMOs (LUMOs) of 2a, 4a, and 6a being a_u (b_{1g}) and those of 3a and 5a being b_{2g} (b_{3u}). Thus, the direct product of the HOMOs (LUMOs) symmetries can be reduced as

$$a_u \times a_u = b_{2g} \times b_{2g} = b_{1g} \times b_{1g} = b_{3u} \times b_{3u} = A_g \tag{24}$$

Therefore, the totally symmetric A_g modes of vibration couple to the HOMOs (LUMOs). The numbers of the A_g vibrational modes are 9, 12, 15, 18, and 21 for **2a**, **3a**, **4a**, **5a**, and **6a**, respectively. The symmetry labels of nondegenerate frontier orbitals of these polyacenes are abbreviated as "*a*" in the following discussion.

The energy sheet of the HOMO is given in the form:

$$\varepsilon_m^{(A_g)}{}_{\text{HOMO}}\left(Q_{A_gm}\right) = \frac{1}{2}K_{A_gm}Q_{A_gm}^2 + C_mQ_{A_gm} \tag{25}$$

where C_m corresponds to the reduced matrix element written as

$$C_m = \left\langle a\,\text{HOMO}\left\|\left(\frac{\partial h}{\partial Q_{A_gm}}\right)_0\right\| a\,\text{HOMO}\right\rangle \tag{26}$$

This is illustrated in Figure 1. The dimensionless diagonal linear orbital vibronic coupling constants of the HOMO for its *m*th mode is defined by Eq. (27),

$$g_{a\,\text{HOMO}}\left(\omega_m\right) = \frac{1}{\hbar\omega_m}\left\langle a\,\text{HOMO}\left\|\left(\frac{\partial h}{\partial q_{A_gm}}\right)_0\right\| a\,\text{HOMO}\right\rangle \tag{27}$$

In this equation, q_{A_gm} is the dimensionless normal coordinate [31] of the *m*th mode defined by using the normal coordinate Q_{A_gm} as

$$q_{A_gm} = \sqrt{\omega_m / \hbar}\,Q_{A_gm} \tag{28}$$

In a similar way, the dimensionless diagonal linear orbital vibronic coupling constant of the LUMO for the *m*th A_g mode is defined by Eq. (29),

$$g_{a\,\text{LUMO}}\left(\omega_m\right) = \frac{1}{\hbar\omega_m}\left\langle a\,\text{LUMO}\left\|\left(\frac{\partial h}{\partial q_{A_gm}}\right)_0\right\| a\,\text{LUMO}\right\rangle \tag{29}$$

D. Vibronic Interactions in Polyfluoroacenes, B, N-Substituted Polyacenes, and Polycyanodienes

Similar discussions can be made in the monocations and anions of polyfluoroacenes, B, N-substituted polyacenes, and polycyanodienes. The numbers of vibronic active modes are 6, 9, 12, 15, 18, and 21 for 1fa, 2fa, 3fa, 4fa, 5fa, and 6fa, respectively, those are 11, 17, and 23 for 1bn, 2bn, and 3bn, respectively, and those are 5, 7, 9, and 11 for 1cn, 2cn, 3cn, and 4cn, respectively.

Chapter III

ELECTRON–PHONON COUPLING CONSTANTS FOR THE CHARGED ELECTRONIC STATES OF POLYACENES, POLYFLUOROACENES, B, N-SUBSTITUTED POLYACENES, AND POLYCYANODIENES

Let us next discuss the total electron–phonon coupling constants (l_{total}) in the monocation and monoanion crystals. Since the l_{total} is the sum of the electron–phonon coupling constants originating from both intramolecular vibrations (l_{intra}) and intermolecular vibrations (l_{inter}), the l_{total} is defined as

$$l_{total} = l_{intra} + l_{inter}.$$

(30)

However, it should be noted that the intramolecular orbital interactions are much stronger than the intermolecular orbital interactions. Therefore, it is rational that the l_{intra} values are much larger than the l_{inter} values in molecular systems. In fact, it is considered by several researchers that the contribution from the intramolecular modes in molecular systems is decisive in the pairing process in the superconductivity in doped C_{60}. [15,32] For example, it was reported that the l_{intra} values are much larger than the l_{inter} values in K_3C_{60} and Rb_3C_{60} ($l_{intra} \geq 10 l_{inter}$). [32] Furthermore, it has also been shown from a neutron-scattering investigation [33] that the electron–libration intramolecular-mode coupling is small in alkali-metal-doped C_{60}. Therefore, we consider only intramolecular electron–phonon coupling in this study. The l_{total} value for the charged and excited electronic states can be defined as

$$l_{total} \approx l_{intra} = l_{LUMO} \text{ (for the electronic state of the monoanion)},$$

$$(31)$$

$$= l_{HOMO} \text{ (for the electronic state of the monocation)},$$

$$(32)$$

In the previous section, the vibronic interactions in free polyacenes, polyfluoroacenes, B, N-substituted polyacenes, and polycyanodienes were discussed. We can derive the l_{total} by using the vibronic coupling constants defined in the previous section as follows. As described above, since polyacenes, polyfluoroacenes, B, N-substituted polyacenes, and polycyanodienes would consist of strongly bonded molecules arranged on a lattice with weak van der Waals intermolecular bonds, we can derive the dimensionless electron–phonon coupling constant λ in a similar way as in theory in previous research. [15,25,27–29] We use a standard expression for λ, [15,25,27–29]

$$\lambda = \frac{2}{N(0)} \sum_{m,q} \sum_{k,k'} \frac{1}{2\omega_{m,q}^2} \left| h_{k,k'}(m,q) \right|^2 \delta(E_k)\delta(E_{k'}),$$

$$(33)$$

where $\omega_{m,q}$ is the vibrational frequency for the mth phonon mode of wave vector q; $h_{k,k'}$ is the corresponding electron–phonon matrix element between the electronic states of wave vectors k and k'; E_k and $E_{k'}$ are the corresponding energies measured from the Fermi level (original point in Figure 1 (b)) of polyacenes, polyfluoroacenes, B, N-substituted polyacenes, and polycyanodienes; $N(0)$ is the total density of states (DOS) at the Fermi level per spin. The charged electronic states in polyacenes, polyfluoroacenes, B, N-substituted polyacenes, and polycyanodienes are essentially composed of the charged electronic states at the equilibrium structures of the ground states in polyacenes, polyfluoroacenes, B, N-substituted polyacenes, and polycyanodienes and we can write in the form of Bloch sum:

$$\psi(k) = \frac{1}{\sqrt{N}} \sum_{R} c(k)e^{ikR}\phi_R$$

$$(34)$$

where R denotes the cell position; N is the number of molecules in the crystal; ϕ_R is wave function which denotes the charged electronic states at the equilibrium structures of the ground state of polyacenes, polyfluoroacenes, B, N-substituted polyacenes, and polycyanodienes at the cell position R. If we neglect the intermolecular electron–lattice coupling, $h_{k,k'}$ can be reduced to

$$h_{k,k'}(m,q) = \langle \psi(k)|h|\psi(k')\rangle \qquad = \frac{1}{N}\sum_R c^*(k)c(k')e^{i(k-k')R} h_{R,R}(m,q)$$

(35)

where $h_{R,R}$ is the intramolecular coupling matrix and

$$h_{R,R}(m,q) = \langle \phi_R|h|\phi_R\rangle$$

(36)

For the one-phonon mode with wave vector q, this term takes the following form:

$$h_{R,R}(m,q) = \left(1/\sqrt{N}\right)e^{iqR} h_{00m}$$

(37)

We insert Eq. (37) in Eq. (35) taking the condition $k' = k - q$ into account to get

$$h_{kk-q} = \frac{1}{\sqrt{N}} h_{00m} c^*(k)c(k-q)$$

(38)

We now proceed to calculate λ by inserting Eq. (38) in Eq. (33) considering that $\omega_{m,q}$ is independent of q. We then obtain

$$\lambda = \frac{2}{N(0)}\sum_{k,k-q}\sum_m \frac{1}{2\omega_{m,q}^2}\frac{h_{00m}h_{00m}{}^*}{N}c^*(k)c(k)c^*(k-q)c(k-q)\delta(E_k)\delta(E_{k-q})$$

(39)

The partial DOS per molecule at the Fermi level of polyacenes, polyfluoroacenes, B, N-substituted polyacenes, and polycyanodienes, $n(0)$ can be rewritten as

$$n(0) = (1/N)\sum_k c*(k)c(k)\delta(E_k)$$

(40)

We can derive Eq. (41) from Eqs. (39) and (40) using $n(0) = N(0)/N$,

$$\lambda = \frac{2}{n(0)}\sum_m \frac{1}{2\omega_m^2} h_{00m} h_{00m}*n(0)^2$$

(41)

where $n(0)$ is now the DOS per spin and per molecule.

A. Twofold Degenerate Electronic States of The Monocations of Monoanions of Benzene, Hexafluorobenzene, and Borazine

Here, for example, let us consider twofold degenerate electronic states such as the monocation of benzene, In such a case, we can assume that $n_{\nu\nu'}(0) = \frac{1}{2}n(0)\delta_{\nu\nu'}$, λ takes the form of Eq. (42):

$$\lambda = \frac{n(0)}{4}\sum_m \frac{h_{00m}^2}{\omega_m^2}$$

(42)

where

$$h_{00m} = h_{E_{2g}m}'$$

(43)

Here $h_{E_{2g}m}'$ is the derivative matrix of the vibronic coupling matrix, $h_{E_{2g}m}$, derived by Eq. (45) with respect to the mode amplitude, $Q_{E_{2g}\gamma m}$, as

$$h_{E_{2g}m}' = \frac{\partial}{\partial Q_{E_{2g}\gamma m}}\left[A_m\begin{pmatrix} -Q_{E_{2g}\varepsilon m} & Q_{E_{2g}\theta m} \\ Q_{E_{2g}\theta m} & Q_{E_{2g}\varepsilon m} \end{pmatrix}\right] = \frac{\partial}{\partial Q_{E_{2g}\gamma m}} h_{E_{2g}m}$$

(44)

where

$$h_{E_{2g}m} = A_m\begin{pmatrix} -Q_{E_{2g}\varepsilon m} & Q_{E_{2g}\theta m} \\ Q_{E_{2g}\theta m} & Q_{E_{2g}\varepsilon m} \end{pmatrix}$$

(45)

and A_m is the reduced matrix element and is the slope in the original point (i.e., equilibrium structures of the ground states of **1a** $(Q_{E_{2g}am} = Q_{E_{2g}am} = 0)$) on the potential energy surface of the charged electronic state along each vibrational mode (Figure 1 (b)), and is defined as

$$A_m = \left\langle e_{1g} \text{ HOMO} \left\| \left(\frac{\partial h}{\partial Q_{E_{2g}m}} \right)_0 \right\| e_{1g} \text{ HOMO} \right\rangle$$

(46)

Since $h_{00m}^2 = 4A_m^2$, one can rewrite Eq. (42) as

$$\lambda = n(0) \sum_m \frac{A_m^2}{\omega_m^2}$$

(47)

Using Eqs. (15) and (32), and after some simple transformations, we finally get the relation between the nondimensional electron–phonon coupling constant λ, and the intramolecular vibronic coupling constant, $g_{e_{1g} \text{ HOMO}}(\omega_m)$, as

$$\lambda = \sum_m \lambda_m$$, $\lambda_m = n(0) l_{e_{1g} \text{ HOMO}}(\omega_m)$,

(48)

where $l_{e_{1g} \text{ HOMO}}(\omega_m)$ is the electron–phonon coupling constant defined as

$$l_{e_{1g} \text{ HOMO}}(\omega_m) = g_{e_{1g} \text{ HOMO}}^2(\omega_m) \hbar\omega_m, \quad (m = 1,2,3,4).$$

(49)

In a similar way, the $l_{e_{1g} \text{ HOMO}}(\omega_m)$ for the A_{1g} vibronic active modes can be defined as

$$l_{e_{1g} \text{ HOMO}}(\omega_m) = \frac{1}{2} g_{e_{1g} \text{ HOMO}}^2(\omega_m) \hbar\omega_m, \quad (m = 5,6).$$

(50)

In a similar way, $l_{e_{2u} \text{ LUMO}}(\omega_m)$ for the monoanion of benzene can be defined as

$$l_{e_{2u} \text{ LUMO}}(\omega_m) = g_{e_{2u} \text{ LUMO}}^2(\omega_m) \hbar\omega_m, \quad (m = 1,2,3,4).$$

(51)

$$l_{e_{2u} \text{LUMO}}(\omega_m) = \frac{1}{2} g_{e_{2u} \text{LUMO}}^2 (\omega_m) \widetilde{\varkappa\omega}_m \quad , (m = 5,6).$$ (52)

Similar discussions can be made in the monocations and monoanions of **1fa** and **1bn**.

B. Nondegenerate Electronic Systems of Polyacenes, Polyfluoroacenes, B, N-Substituted Polyacenes, and Polycyanodienes

In a similar way, by using Eq. (41), the $l_{\text{HOMO}}(\omega_m)$ and $l_{\text{LUMO}}(\omega_m)$ for the nondegenerate electronic states in the monocations and monoanions of polyacenes, polyfluoroacenes, B, N-substituted polyacenes, and polycyanodienes can be defined as

$$l_{\text{HOMO}}(\omega_m) = g_{\text{HOMO}}^2 (\omega_m) \widetilde{\varkappa\omega}_m ,$$ (53)

$$l_{\text{LUMO}}(\omega_m) = g_{\text{LUMO}}^2 (\omega_m) \widetilde{\varkappa\omega}_m .$$ (54)

Chapter IV

OPTIMIZED STRUCTURES

A. Benzene and Polyacenes

The structures of neutral 1a and polyacenes were optimized under D_{6h} and D_{2h} symmetries, respectively, using the hybrid Hartree–Fock (HF)/density-functional-theory (DFT) method of Becke [34] and Lee, Yang, and Parr [35] (B3LYP) and the 6-31G* basis set. [36] GAUSSIAN 98 program package [37] was used for our theoretical analyses. This level of theory is, in our experience, sufficient for reasonable descriptions of the geometric, electronic, and vibrational structures of hydrocarbons. An optimized D_{6h} structure of 1a and D_{2h} structures of polyacenes are shown in Figure 2. These structures were confirmed to be a minimum on each potential energy surface from vibrational analyses. We can see that in the D_{6h} structure of 1a there is no bond alternation and all C–C bond lengths are approximately 1.4 Å. In the D_{2h} structure of 2a, there is a distinct variation in the C–C distances. This result can be understood in view of the orbital patterns of the HOMO. Selected vibronic active modes and the phase patterns of the HOMOs and LUMOs of the neutral molecules under consideration are shown in Figure 3. We can see from this figure that the atomic orbitals between two neighboring C_{2a} and C_{3a} atoms are combined in phase and thus form a strong π-bonding in the HOMO of 2a. On the other hand, the HOMO contributes nonbonding interactions between C_{1a} and C_{1b} atoms and between C_{1a} and C_{2a} atoms, and an antibonding interaction between C_{3a} and C_{3b} atoms. This is the reason why the C_{2a}–C_{3a} bond is much shorter (1.377 Å) than any other C–C bond (1.417–1.434 Å). Similar discussions can be made in 3a, 4a, 5a, and 6a; the C–C bond distances between two neighboring carbon atoms whose atomic orbitals are combined in phase (out of phase) are short (long).

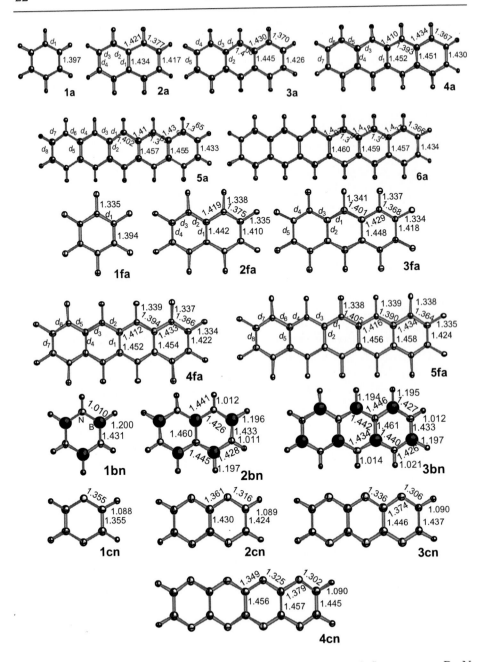

Figure 2. Optimized structures of the neutral polyacenes, polyfluoroacenes, B, N-substituted polyacenes, and polycyanodienes.

B. Polyfluoroacenes

Hexafluorobenzene (C_6F_6) (1fa) has been well studied for a long time. [38–53] 1a and 1fa have nearly the same polarizations and polarizability anisotropies. [50] As well as being similar in molecular geometry and shape, 1a and 1fa have similar physical properties. [51] Their electric quadruple moments are roughly equal in magnitude but opposite in sign. [52,53] In recent review, [54] the study of Jahn–Teller and pseudo-Jahn–Teller effects in various molecular systems including the monoanion [55] and cation [56,57] of 1fa are well summarized.

The structures of neutral 1fa was optimized under D_{6h} symmetry, and 2fa, 3fa, 4fa, and 5fa were optimized under D_{2h} symmetry. According to our calculations, the energy differences between the HOMO and LUMO of 1fa, 2fa, 3fa, 4fa, and 5fa are 6.52, 4.56, 3.35, 2.57, and 2.03 eV, respectively, and those of 1a, 2a, 3a, 4a, and 5a are 6.80, 4.83, 3.59, 2.78, and 2.21 eV, respectively. Therefore, the HOMO–LUMO gaps in polyfluoroacenes are slightly smaller than those in polyacenes. This can be understood in view of the orbital patterns of the HOMO and LUMO in these molecules. We can see from Figure 3 that the phase patterns of the LUMO of polyfluoroacenes are not significantly different from those of polyacenes. The phase patterns of the LUMO in carbon framework of polyfluoroacenes are similar to those of polyacenes in that the LUMO of polyfluoroacenes is rather localized on carbon framework and the electron density on fluorine atoms is low, and the LUMO of polyacenes is completely localized on carbon atoms. The phase patterns of the HOMO in carbon framework of polyfluoroacenes are also similar to those of polyacenes. However, the HOMO of polyfluoroacenes is delocalized and the electron density on fluorine atoms is not small, while the HOMO of polyacenes is completely localized on carbon atoms. The atomic orbitals between two neighboring carbon and fluorine atoms are combined out of phase in the HOMO in polyfluoroacenes. Therefore, the HOMO is slightly destabilized in energy measured from the LUMO in polyfluoroacenes by H–F substitution. This is the reason why the energy difference between the HOMO and LUMO in polyfluoroacenes is slightly smaller than that in polyacenes.

Let us look into optimized structures of polyfluoroacenes. We can see from Figure 2 that the calculated C–C distances in 1fa are 1.394 Å. There is a distinct variation in the C–C distances in polyfluoroacenes. This result is reasonable in view of the HOMOs. The C–C distances between two neighboring carbon atoms whose atomic orbitals are combined in phase (out

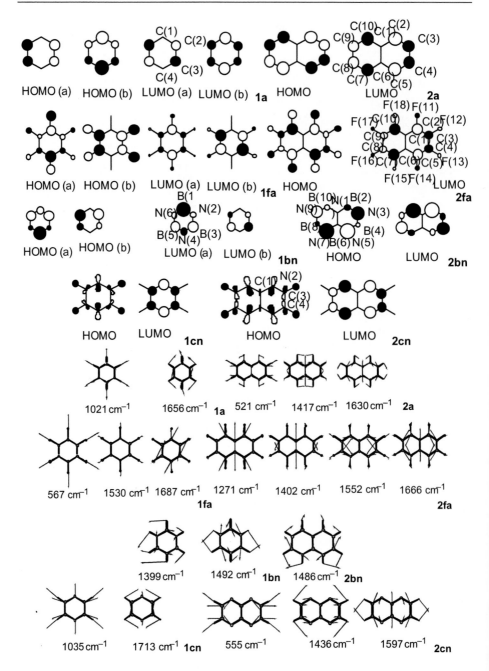

Figure 3. The phase patterns of the HOMO and LUMO and selected vibronic active modes in polyacenes, polyfluoroacenes, B, N-substituted polyacenes, and polycyanodienes.

of phase) in the HOMOs of polyfluoroacenes are short (long). It should be noted that the C–C distances in polyfluoroacenes are similar to those in polyacenes; most of the C–C bond lengths differences between polyacenes and polyfluoroacenes are smaller than 0.003 Å. Only the d_4 value difference between 2a and 2fa, d_5 value difference between 3a and 3fa, d_7 value difference between 4a and 4fa, and d_8 value difference between 5a and 5fa are somewhat large (0.007–0.009 Å). Therefore, carbon framework structures in polyfluoroacenes are not significantly different from those in polyacenes as a consequence of H–F substitution.

C. B, N-Substituted Polyacenes

Substitution of carbon with boron and nitrogen have been carried out, producing totally substituted boron nitride nanotubes (BN-NTs). [58] Many researchers have studied and compared the structure of these new boron–nitrogen-containing fullerenes and nanotubes with that of carbon nanotubes (C-NTs). [59–73] A large number of molecular boron–nitrogen compounds have been prepared. [74–79] Because borazine ($B_3N_3H_6$) (1bn) is both isoelectronic and isosteric with 1a, it is a very interesting compound. Stock and Pohland [80] first reported formation of borazine 1bn from the thermolysis (200 °C) of the addition complex $[H_2B(NH_2)_2{}^+][BH_4{}^-]$, [81] formulated in 1926 as $B_2N_6 \cdot 2NH_3$. Laubengayer and co-workers [82] studied the gas-phase pyrolysis of 1bn in greater detail, and they proposed the intermediate formation of a naphthalene analogue, $B_5N_5H_8$ (2bn).

The structure of neutral "inorganic benzene", $B_3N_3H_6$ (1bn) was optimized under D_{3h} symmetry, and "inorganic-" naphthalene ($B_5N_5H_8$) (2bn) and anthracene ($B_7N_7H_{10}$) (3bn) were optimized under C_{2v} symmetry. According to our calculations, the energy difference between the HOMO and the LUMO of 1bn (8.27 eV) is larger than that of 1a (6.80 eV). Furthermore, that of 2bn (7.12 eV) is larger than that of 2a (4.83 eV), and that of 3bn (6.68 eV) is larger than that of 3a (3.59 eV). This can be understood as follows. The degenerate 1a levels have not been split apart in energy but the HOMO contains more nitrogen character than boron character while opposite is true for the LUMO.[30] This is a result clearly in keeping with an electronegativity perturbation on 1a. [30] Due to the electronegativity perturbation, the HOMO (LUMO) of 1bn is stabilized (destabilized) in energy with respect to the HOMO (LUMO) of 1a. This is the reason why the HOMO–LUMO gap is larger in 1bn than in 1a. Similar discussions can be made in 2bn and 3bn. Let us next look into optimized structure of 1bn. We can see from Figure 2 that the calculated B–N, B–H, and N–H distances in 1bn are 1.431, 1.200, and 1.010 Å, respectively, and those determined from experimental method [83] are 1.44, 1.20, and 1.02 Å, respectively. Therefore, the structure optimized at the B3LYP/6-31G* level of theory is in excellent agreement with that determined from the experimental method. There is a distinct variation in the B–N distances in 2bn and 3bn. This result is reasonable in view of the HOMOs shown in Figure 3. The B–N distances between two neighboring B and N atoms whose atomic orbitals are combined in phase (out of phase) in the HOMOs of 2bn and 3bn are short (long). It should be noted that the B–N distance (1.431 Å) in 1bn is larger than the C–C distance (1.397 Å) in 1a, but smaller than the B–N single bond distance (1.60 Å). Therefore, it can be said that each B–N bond in 1bn acquires double bond character. Similar discussion can be made in 2bn and 3bn; all B–N bonds in 2bn and 3bn are longer than all C–C bonds in 2a and 3a, but smaller than 1.6 Å.

D. Polycyanodienes

We can expect that in the monocations of polycyanodienes, the electron–phonon interactions become much stronger than in the monocations of polyacenes because the strong σ-orbital interactions as well as the π-orbital interactions between two neighboring atoms in the HOMO are significant, [84] and because the HOMO is delocalized and the electron density on hydrogen atoms as well as on carbon and nitrogen atoms is high in the HOMO of polycyanodienes. We call this system "σ-conjugated momocations".

The structures of polycyanodienes were optimized under D_{2h} symmetry. According to our calculations, the energy differences between the HOMO and LUMO of 1cn, 2cn, 3cn, and 4cn are 5.16, 3.97, 3.19, and 2.68 eV, respectively. Therefore, the HOMO–LUMO gaps in polycyanodienes are slightly smaller than those in polyacenes.

Chapter V

ELECTRON–PHONON COUPLING CONSTANTS

A. The Monocations and Monoanions of Benzene and Polyacenes

We next carried out vibrational analyses of 1a and polyacenes at the B3LYP/6-31G* level. Agreement with experiment is excellent for the vibrational frequencies, calculated wave numbers of 1a being accurate within a range of 2.5–4 %. Since the conventional Hartree–Fock method overestimates molecular vibrational frequencies by typically 10 %, this hybrid DFT method appears to be very useful for vibrational analyses. We next calculated first-order derivatives at this equilibrium structure on each orbital energy surface by distorting the molecule along the A_{1g} and E_{2g} modes of 1a and the A_g modes of polyacenes. In these calculations, the step size of the normal-mode displacements was taken to be in the order of 10^{-1}. What we obtained from the first-order derivatives are the dimensionless diagonal linear orbital vibronic coupling constants $g_{HOMO}(\omega_m)$ and $g_{LUMO}(\omega_m)$. We can estimate the electron–phonon coupling constants $l_{HOMO}(\omega_m)$ and $l_{LUMO}(\omega_m)$ from the dimensionless diagonal linear orbital vibronic coupling constants by using Eqs. (49)–(54). The calculated electron–phonon coupling constants in the monocatins and anions of 1a and polyacenes are shown in Figures 4 and 5, respectively.

Let us next take a look at the electron–phonon coupling of the A_{1g} and E_{2g} vibrational modes to the e_{1g} HOMO in 1a. E_{2g} modes of 1a are classified into a C–C–C in-plane bending (622 cm^{-1}), a C–H in-plane bending (1208 cm^{-1}), a C–C stretching (1656 cm^{-1}), and a C–H stretching (3184 cm^{-1}). Figure 4 demonstrates that the E_{2g} vibrational modes of 622 and 1656 cm^{-1} strongly couple to the e_{1g} HOMO in 1a. This can be understood in view of the orbital patterns of the e_{2u} LUMO and vibrational modes of 1a. When 1a is distorted along this E_{2g} mode toward the direction as shown in Figure 3, the antibonding (bonding) interactions between C_{1b} and C_{1c} in the HOMO (a) (HOMO (b)) become strong, and the bonding interactions between C_{1a} and C_{1b} and between C_{1c} and C_{1d} in the HOMO (a) become weak. Therefore, the HOMO (a) (HOMO (b)) is significantly destabilized (stabilized) in energy. On the other hand, when 1a is distorted toward the opposite direction along the arrow of this mode, the HOMO (a) (HOMO (b)) is significantly stabilized (destabilized) in energy. This is the reason why the E_{2g} mode of 1656 cm^{-1} strongly couples to the e_{1g} HOMO in 1a. Figure 4 demonstrates that the C–C stretching A_g modes of 1417 and 1630 cm^{-1} strongly couple to the HOMO in 2a. When 2a is distorted along the A_g mode of 1417 cm^{-1} toward the direction as shown in Figure 3, the antibonding interactions between C_{3a} and C_{3b} in the HOMO become strong, and the bonding interactions between C_{2a} and C_{3a} and between C_{2b} and C_{3b} become weak.

Therefore, the HOMO of 2a is significantly destabilized in energy. On the other hand, when 2a is distorted toward the opposite direction, the HOMO is significantly stabilized in energy. This is the reason why the C–C stretching A_g mode of 1417 cm^{-1} strongly couples to the HOMO. In a similar way, the C–C stretching A_g mode of 1630 cm^{-1} strongly couples to the HOMO. In the two frequency modes, the displacements of carbon atoms are very large compared with any other A_g mode of 2a, and the HOMO is completely localized on carbon atoms. It is rational that a frequency mode in which normal displacements are large for atoms where there exists considerable orbital amplitude can strongly couple to the molecular orbital. Identical discussions can be made in 3a, 4a, 5a, and 6a; the C–C stretching A_g modes around 1500 cm^{-1} and the lowest frequency A_g mode strongly couple to the HOMOs.

Let us take a look at the electron–phonon coupling of the A_{1g} and E_{2g} vibrational modes to the e_{2u} LUMO in 1a. We can see from Figure 5 that the E_{2g} mode of 1656 cm^{-1} strongly couples to the e_{2u} LUMO in 1a. This can be also understood in view of the phase patterns of the e_{2u} LUMO and vibrational modes of 1a. Figure 5 demonstrates that the C–C stretching A_g modes of 1417 and 1630 cm^{-1} strongly couple to the b_{1g} LUMO. In the two frequency modes, the displacements of carbon atoms are very large compared with any other A_g mode of 2a, and the b_{1g} LUMO is completely localized on carbon atoms. It is rational that a frequency mode in which normal displacements are large for atoms where

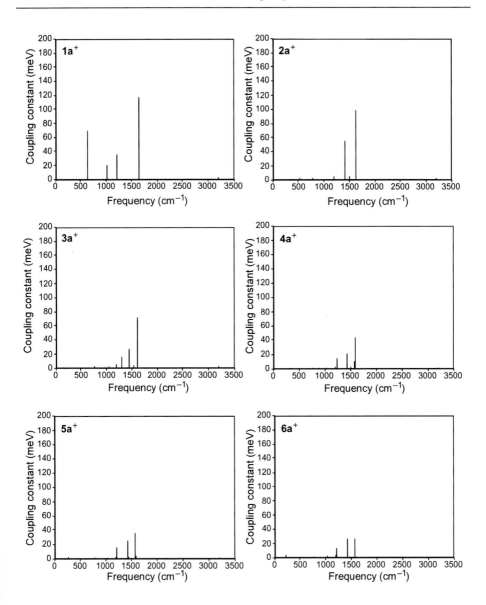

Figure 4. Electron–phonon coupling constants for the monocations of polyacenes.

there exists considerable orbital amplitude can strongly couple to the molecular orbital. Identical discussions can be made in 3a, 4a, 5a, and 6a; the C–C stretching A_g modes around 1500 cm^{-1} and the lowest frequency A_g mode strongly couple to the LUMOs. The C–C stretching A_g modes around 1500 cm^{-1} strongly couple to the HOMOs as well as to the LUMOs in polyacenes. But it should be noted that the lowest frequency modes as well as the C–C stretching modes strongly couple to the LUMO in polyacenes, while the only C–C stretching modes strongly couple to the HOMO in polyacenes.

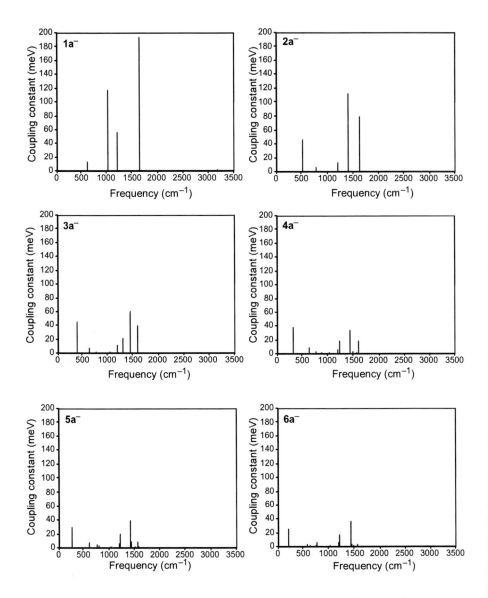

Figure 5. Electron–phonon coupling constants for the monoanions of polyacenes.

B. Polyfluoroacenes

1. Monocations

We carried out vibrational analyses of polyfluoroacenes at the B3LYP/6-31G* level of theory. There are four E_{2g} modes in 1fa (C–C–F in-plane bending mode of 259 cm^{-1}; C–C–C in-plane bending mode of 444 cm^{-1}; C–F stretching mode of 1192 cm^{-1}; C–C stretching mode of 1687 cm^{-1}). There are four E_{2g} vibrational modes in 1a (C–C–C in-plane bending mode of 622 cm^{-1}; C–C–H in-plane bending mode of 1208 cm^{-1}; C–C stretching mode of 1656 cm^{-1}; C–H stretching mode of 3184 cm^{-1}). Therefore, the frequencies for the C–C–H in-plane bending mode (1208 cm^{-1}) and for the C–H stretching mode (3184 cm^{-1}) in 1a are much larger than those for the C–C–F in-plane bending mode (259 cm^{-1}) and for the C–F stretching mode (1192 cm^{-1}) in 1fa, respectively, as expected. Furthermore, the frequency for the C–C–C in-plane bending mode (622 cm^{-1}) in 1a is larger than that (444 cm^{-1}) in 1fa. The frequency for the C–C stretching mode (1656 cm^{-1}) in 1a is similar to that (1687 cm^{-1}) in 1fa.

The calculated electron–phonon coupling constants in the monocations and monoanions of 1fa, 2fa, 3fa, 4fa, and 5fa are shown in Figures 6 and 7, respectively. We can see from Figure 6 that the C–C stretching E_{2g} mode of 1687 cm^{-1} can strongly couple to the e_{1g} HOMO in 1fa. This can be understood in view of the phase patterns of the HOMO in 1fa. The reduced mass for the E_{2g} mode of 1687 cm^{-1} is 12.05, and the displacements of carbon atoms are much larger than those of fluorine atoms in the mode.

When 1fa is distorted along the E_{2g} mode of 1687 cm^{-1} toward the same direction as shown in Figure 3, the antibonding (bonding) interactions between two neighboring carbon atoms in the HOMO (a) (HOMO (b)) become significantly stronger, and the bonding interactions between two neighboring carbon atoms in the HOMO (a) become significantly weaker, and thus the HOMO (a) (HOMO (b)) is significantly destabilized (stabilized) in energy by such a distortion. This is the reason why the E_{2g} mode of 1687 cm^{-1} the most strongly couples to the e_{1g} HOMO in 1fa.

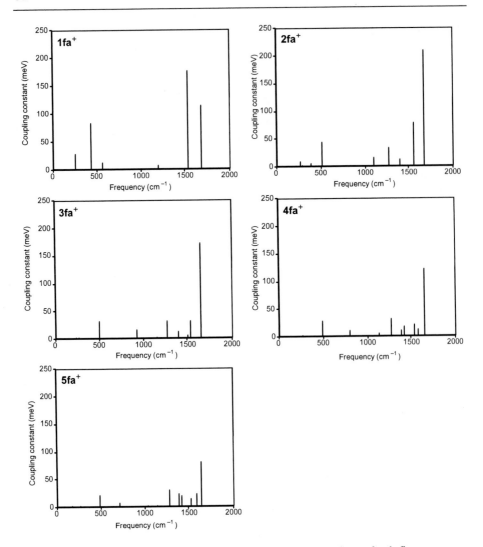

Figure 6. Electron–phonon coupling constants for the monocations of polyfluoroacenes.

We can see from Figure 6 that the A_{1g} mode of 1530 cm^{-1} can very strongly couple to the e_{1g} HOMO in 1fa. This can be understood as follows. When 1fa is distorted along the A_{1g} mode of 1530 cm^{-1}, the bonding interactions between two neighboring carbon atoms become significantly weaker, and the antibonding interactions between two neighboring carbon and fluorine atoms in the e_{1g} HOMO become significantly stronger, and thus the e_{1g} HOMO is significantly

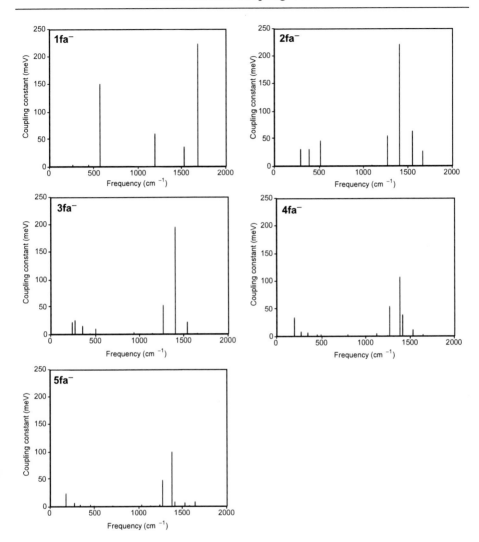

Figure 7. Electron–phonon coupling constants for the monoanions of polyfluoroacenes.

destabilized in energy by such a distortion in 1fa. This is the reason why the A_{1g} mode of 1530 cm^{-1} can strongly couple to the e_{1g} HOMO in 1fa. The A_{1g} mode of 567 cm^{-1} much less strongly couples to the e_{1g} HOMO than the A_{1g} mode of 1530 cm^{-1} in 1fa. This can be understood as follows. When 1fa is distorted along the A_{1g} mode of 567 cm^{-1}, the bonding interactions between two neighboring carbon atoms and the antibonding interactions between two neighboring carbon and fluorine atoms in the e_{1g} HOMO become weaker. Since such weakened effects of bonding and antibonding interactions are compensated by each other, the e_{1g} HOMO is slightly stabilized in energy by such a distortion in 1fa. This is the reason why the A_{1g} mode of 567 cm^{-1} much less strongly couples to the e_{1g} HOMO than the A_{1g} mode of 1530 cm^{-1} in 1fa.

We can see from Figure 6 that the C–C stretching A_g mode of 1666 cm^{-1} the most strongly couples to the HOMO in 2fa. This can be understood as follows. When 2fa is distorted along the A_g mode of 1666 cm^{-1}, the bonding and antibonding interactions between two neighboring carbon atoms become significantly stronger and weaker, respectively, and the antibonding interactions between two neighboring carbon and fluorine atoms in the HOMO in 2fa become weaker. Therefore, the HOMO is significantly stabilized in energy by such a distortion in 2fa. This is the reason why the A_g mode of 1666 cm^{-1} can the most strongly couple to the HOMO in 2fa. In a similar way, the A_g modes of 514 and 1552 cm^{-1} also strongly couple to the HOMO in 2fa. Similar discussions can be made in 3fa, 4fa, and 5fa; the low frequency mode around 500 cm^{-1}, the high frequency modes around 1200 cm^{-1}, and the frequency modes around 1600 cm^{-1} afford large electron–phonon coupling constants in the monocations of polyfluoroacenes.

2. Monoanions

Let us first look into the electron–phonon interactions between the E_{2g} modes and the e_{2u} LUMO in 1fa. We can see from this figure that the C–C stretching E_{2g} mode of 1687 cm^{-1} can strongly couple to the e_{2u} LUMO in 1fa. This can be understood in view of the phase patterns of the LUMO in 1fa. When 1fa is distorted along the E_{2g} mode of 1687 cm^{-1} toward the same direction as shown in Figure 3, the bonding (antibonding) interactions between two neighboring carbon atoms in the LUMO (a) become stronger (weaker), therefore, the LUMO (a) is significantly stabilized in energy. In a similar way, the antibonding interactions between two neighboring carbon atoms in the LUMO (b) become stronger by such a distortion, therefore, the LUMO (b) is significantly destabilized in energy. This is the reason why the E_{2g} mode of 1687 cm^{-1} can strongly couple to the e_{2u} LUMO in 1fa. In addition to this mode, the C–F stretching E_{2g} mode of 1192 cm^{-1} can strongly couple to the e_{2u} LUMO, but the E_{2g} modes of 259 and 444 cm^{-1} hardly couple to it. This can be understood as follows. The reduced masses for the E_{2g} modes of 1192 and 1687 cm^{-1} are 13.69 and 12.05, respectively, therefore, the displacements of carbon atoms are much larger than those of fluorine atoms in these modes. On the other hand, the reduced masses for the E_{2g} modes of 259 and 444 cm^{-1} are 18.62 and 16.08, respectively, and therefore, the displacements of fluorine atoms are more dominant than those of carbon atoms in these modes. Since the e_{2u} LUMO is rather localized on carbon atoms, it is reasonable that a frequency mode in which normal displacements of carbon atoms are large (small), can strongly (weakly) couple to the e_{2u} LUMO. This is the reason why the E_{2g} modes of 1192 and 1687 cm^{-1} can strongly couple to the e_{2u} LUMO, while the E_{2g} modes of 259 and 444 cm^{-1} hardly couple to it.

Let us next look into the electron–phonon interactions between the A_{1g} modes and the e_{2u} LUMO in 1fa. We can see from Figure 7 that the A_{1g} mode of 567 cm^{-1} much more strongly couples to the e_{2u} LUMO than the A_{1g} mode of 1530 cm^{-1} in 1fa. This can be understood as follows. When 1fa is distorted along the A_{1g} mode of 567 cm^{-1}, the antibonding interactions between two neighboring carbon atoms, and between two neighboring carbon and fluorine atoms, both in the LUMO (a) and LUMO (b), become weaker. Therefore, both the LUMO (a) and LUMO (b) are significantly stabilized in energy by such a distortion. When 1fa is distorted along the A_{1g} mode of 1530 cm^{-1}, the antibonding interactions between two neighboring carbon atoms, and between two neighboring carbon and fluorine atoms, become weaker and stronger, respectively, both in the LUMO (a) and LUMO (b). Since such strengthened and weakened effects of antibonding interactions are compensated by each other, the A_{1g} mode of 1530 cm^{-1} less strongly couples to the e_{2u} LUMO than the A_{1g} mode of 567 cm^{-1}.

Let us next look into the electron–phonon interactions between the A_g modes and the b_{1g} LUMO in 2fa. We can see from Figure 7 that the C–C stretching A_g mode of 1402 cm^{-1} much more strongly couples to the LUMO than the other modes in 2fa. This can be understood as follows. The reduced mass for the A_g mode of 1402 cm^{-1} is 12.11, therefore, the displacements of carbon atoms are much larger than those of fluorine atoms in the mode. When 2fa is distorted along the A_g mode of 1402 cm^{-1}, the bonding (antibonding) interactions between two neighboring carbon atoms become stronger (weaker), and the antibonding interactions between two neighboring carbon and fluorine atoms in the LUMO become weaker. Therefore, the LUMO is significantly stabilized in energy by such a distortion. This is the reason why the A_g mode of 1402 cm^{-1} strongly couples to the LUMO in 2fa. We can see from Figure 7 that the A_g modes of 1271 and 1552 cm^{-1} also somewhat strongly couple to the LUMO in 2fa. This can be understood as follows. When 2fa is distorted along the A_g mode of 1552 cm^{-1}, the bonding interactions between C(3) and C(4) atoms, and between C(8) and C(9) atoms, become stronger, and the antibonding interactions between C(3) and F(12) atoms, between C(4) and F(13) atoms, between C(8) and F(16) atoms, and between C(9) and F(17) atoms, in the LUMO, become weaker. Therefore, the LUMO is stabilized in energy by such a distortion. This is the reason why the A_g mode of 1552 cm^{-1} also somewhat strongly couples to the LUMO in 2fa. But it should be noted that the antibonding interactions between C(2) and C(3) atoms, between C(4) and C(5) atoms, between C(7) and C(8) atoms, and between C(9) and C(10) atoms in the LUMO, become slightly stronger, and thus the LUMO would be slightly destabilized in energy by such an effect. The destabilization effect and the stabilization effect described above are compensated by each other, the LUMO is slightly stabilized in energy by such a distortion. This is the reason why the A_g mode of 1552 cm^{-1} much less strongly couples to the LUMO than the A_g mode of 1402 cm^{-1} even though both A_g modes are C–C stretching modes and the reduced mass for the A_g mode of 1552 cm^{-1} is similar to that for the A_g mode of 1402 cm^{-1}. When 2fa is distorted along the C–F stretching A_g mode of 1271 cm^{-1}, the antibonding interactions between C(2) and F(11) atoms, between C(5) and F(14) atoms, between C(7) and F(15) atoms, and between C(10) and F(18) atoms in the LUMO, become stronger, therefore, the LUMO is destabilized in energy by such a distortion. But there are not significant changes in the strengths of the orbital interactions between two neighboring carbon atoms in the mode, and thus the C–F stretching A_g mode of 1271 cm^{-1} much less strongly couples to the LUMO than the C–C stretching A_g mode of 1402 cm^{-1} in 2fa. Similar discussions can be made in 3fa, 4fa, and 5fa; the C–C stretching A_g modes of 1401, 1389, and 1379 cm^{-1} the most strongly couple to the LUMO in 3fa, 4fa, and 5fa, respectively.

Let us next compare the calculated results in polyfluoroacenes with those in polyacenes. As in case of the monoanion of 1fa, the C–C stretching E_{2g} mode of 1656 cm^{-1} affords the largest electron–phonon coupling constant (193 meV) in the monoanion of 1a. But it should be noted that the C–C stretching E_{2g} mode of 1687 cm^{-1} in the monoanion of 1fa affords a larger electron–phonon coupling constant (223 meV) than the C–C stretching E_{2g} mode of 1656 cm^{-1} in the monoanion of 1a. This can be understood as follows. The reduced mass for the E_{2g} mode of 1656 cm^{-1} in 1a is 5.32, and therefore, the displacements of hydrogen atoms as well as those of carbon atoms are large in the mode. On the other hand, the reduced mass for the E_{2g} mode of 1687 cm^{-1} in 1fa is 12.05, and thus, the displacements of carbon atoms are very large and those of fluorine atoms are very small in the mode. The LUMO is completely localized on carbon atoms in 1a, and is rather localized on carbon atoms in 1fa. Therefore, it is rational that the E_{2g} mode of 1687 cm^{-1} in 1fa in which the displacements of carbon atoms are larger, more strongly couples to the LUMO localized on carbon atoms than the E_{2g} mode of 1656 cm^{-1} in 1a in which the displacements of carbon atoms are smaller.

Similar discussions can be made in 2fa, 3fa, 4fa, and 5fa; the C–C stretching modes afford larger electron–phonon coupling constants in the monoanions of 2fa, 3fa, 4fa, and 5fa than in the monoanions of 2a, 3a, 4a, and 5a, respectively. This is because the displacements of carbon atoms in the modes are larger in 2fa, 3fa, 4fa, and 5fa than those in 2a, 3a, 4a, and 5a, respectively, and the orbital patterns of the LUMO in polyacenes hardly change by H–F substitution.

It should be noted that the C–C stretching modes around 1500 cm^{-1} and the low frequency modes, less and more, respectively, strongly couple to the LUMO with an increase in molecular size in polyacenes, while the low frequency modes as well as the C–C stretching modes around 1500 cm^{-1} less strongly couple to the LUMO with an increase in molecular size in polyfluoroacenes.

It should be noted that the A_{1g} mode of 567 cm^{-1} (1530 cm^{-1}) much more (less) strongly couples to the e_{2u} LUMO than to the e_{1g} HOMO in 1fa. This can be understood as follows. When 1fa is distorted along the A_{1g} mode of 567 cm^{-1}, the antibonding interactions between two neighboring carbon atoms and between two neighboring carbon and fluorine atoms in the e_{2u} LUMO in 1fa become significantly weaker, and thus the e_{2u} LUMO is significantly stabilized in energy, while the energy level of the e_{1g} HOMO does not significantly change by such a distortion because the weakened effects of bonding and antibonding interactions are compensated by each other, as described above. This is the reason why the A_{1g} mode of 567 cm^{-1} much more strongly couples to the e_{2u} LUMO than to the e_{1g} HOMO in 1fa. When 1fa is distorted along the A_{1g} mode of 1530 cm^{-1}, the antibonding interactions between two neighboring carbon atoms and between two neighboring carbon and fluorine atoms in the e_{2u} LUMO in 1fa become weaker and stronger, respectively. Since such weakened and strengthened effects of antibonding interactions are compensated by each other, the e_{2u} LUMO in 1fa is slightly destabilized in energy by such a distortion. On the other hand, such a compensation does not occur in the electron–phonon interactions between the e_{1g} HOMO and the A_{1g} mode of 1530 cm^{-1}, as described above. This is the reason why the A_{1g} mode of 1530 cm^{-1} much more strongly couples to the e_{1g} HOMO than to the e_{2u} LUMO in 1fa. But it should be noted that the high frequency E_{2g} modes of 1192 and 1687 cm^{-1} (the low frequency E_{2g} modes of 259 and 444 cm^{-1}) much more (less) strongly couple to the e_{2u} LUMO than to the e_{1g} HOMO in 1fa. This can be understood as follows. The reduced masses for the high frequency E_{2g} modes of 1192 and 1687 cm^{-1} are 13.69 and 12.05, respectively, and thus the displacements of carbon atoms are much larger than those of fluorine atoms in these modes in 1fa, while those for the low frequency E_{2g} modes of 259 and 444 cm^{-1} are 18.62 and 16.08, respectively, and thus the displacements of fluorine atoms are larger than those of carbon atoms in these modes in 1fa. The e_{2u} LUMO is rather localized on carbon atoms and in which the electron density on fluorine atoms are low, on the other hand, the e_{1g} HOMO is delocalized and in which the electron density on fluorine atoms are as high as those on carbon atoms in 1fa. It is rational that the higher (lower) frequency E_{2g} modes, in which the displacements of carbon (fluorine) atoms are very large, more (less) strongly couple to the e_{2u} LUMO rather localized on carbon atoms than to the delocalized e_{1g} HOMO in 1fa. This is the reason why the high frequency E_{2g} modes of 1192 and 1687 cm^{-1} (the low frequency E_{2g} modes of 259 and 444 cm^{-1}) much more (less) strongly couple to the e_{2u} LUMO than to the e_{1g} HOMO in 1fa.

It should be noted that the A_g mode of 1666 cm^{-1} (1402 cm^{-1}) much more (less) strongly couples to the HOMO than to the LUMO in 2fa. This can be understood as follows. When 2fa is distorted along the A_g mode of 1666 cm^{-1}, the bonding and antibonding interactions between two neighboring carbon atoms in the LUMO in 2fa become weaker and stronger, respectively, and thus the LUMO would be destabilized in energy by such a distortion. On the other hand, the antibonding interactions between two neighboring carbon and fluorine atoms in the LUMO in 2fa become weaker, and thus the LUMO would be stabilized in energy by such a distortion. Since such destabilization and stabilization effects are compensated by each other, the LUMO is slightly destabilized in energy by such a distortion in 2fa. Such a compensation does not occur in the electron–phonon interactions between the A_g mode of 1666 cm^{-1} and the HOMO in 2fa, and thus the A_g mode of 1666 cm^{-1} much more strongly couples to the HOMO than to the LUMO in 2fa. In a similar way, such a compensation occurs (does not occur) in the electron–phonon interactions between the A_g mode of 1402 cm^{-1} and the HOMO (LUMO), and thus the energy level of the LUMO much more significantly changes than that of the HOMO by such a distortion in 2fa. This is the reason why the A_g mode of 1402 cm^{-1} can much more strongly couple to the LUMO than to the HOMO in 2fa. That is, the significant phase patterns difference between the HOMO, in which the atomic orbitals between two neighboring carbon atoms are mainly combined in phase and those between two neighboring carbon and fluorine atoms are combined out of phase, and the LUMO, in which the atomic orbitals between two neighboring carbon atoms and between two neighboring carbon and fluorine atoms are combined out of phase, is the main reason why the vibrational modes which play an essential role in the electron–phonon interactions in the monocations are significantly different from those in the monoanions in 2fa.

In a similar way, we can rationalize in view of the phase patterns of the HOMO and LUMO the calculated results that the frequency modes lower than 500 cm^{-1} and the high frequency modes around 1400 cm^{-1} much more strongly couple to the LUMO than to the HOMO in 2fa, 3fa, 4fa, and 5fa, on the other hand, the frequency modes around 500 cm^{-1} and the frequency modes around 1600 cm^{-1} much more strongly couple to the HOMO than to the LUMO in 2fa, 3fa, 4fa, and 5fa.

C. B, N-Substituted Polyacenes

1. Monocations

There are five kinds of E' vibrational modes in 1bn (B–N–B or N–B–N in-plane bending mode of 520 cm^{-1}; B–N–H or N–B–H in-plane bending modes of 951 and 1073 cm^{-1}; B–N stretching modes of 1399 and 1492 cm^{-1}; B–H stretching mode of 2625 cm^{-1}; N–H stretching mode of 3640 cm^{-1}). The frequencies of the vibrational modes in B, N-substituted polyacene-series are slightly lower than those in polyacene-series. For example, the frequencies of the B–N stretching modes of 1399 and 1492 cm^{-1} in 1bn are lower than that of the C–C stretching mode of 1656 cm^{-1} in 1a. This can be understood as follows. The B–N bonds in 1bn are longer than the C–C bonds in 1a. In general, the frequency of the stretching mode between two neighboring atoms whose bond lengths are larger, is lower. The C–H (1.087 Å) bond lengths in 1a lie between the N–H (1.010 Å) and B–H (1.200 Å) bond lengths in 1bn, and the frequency for the C–H (3184 cm^{-1}) stretching mode in 1a also lies between those for the N–H (3640 cm^{-1}) and B–H (2625 cm^{-1}) stretching modes in 1bn.

The calculated electron–phonon coupling constants in the monocations and monoanions of 1bn, 2bn, and 3bn are shown in Figure 8. We can see from Figure 3 that the HOMO and LUMO in B, N-substituted polyacene-series are rather localized on N and B atoms, respectively, due to the electronegativity perturbation on polyacene-series. We can see from Figure 8 that the B–N stretching E' mode of 1492 cm^{-1} can strongly couple to the e' HOMO in 1bn. This can be understood in view of phase patterns of the HOMO in 1bn. When 1bn is distorted along the E' mode of 1492 cm^{-1} toward the same direction as shown in Figure 3, the bonding (antibonding) interactions between two neighboring B and N atoms in the HOMO (a) become stronger (weaker), therefore, the HOMO (a) is significantly stabilized in energy. In a similar way, the bonding interactions between two neighboring B and N atoms in the HOMO (b) become weaker by such a distortion, therefore, the HOMO (b) is significantly destabilized in energy. This is the reason why the E' mode of 1492 cm^{-1} can strongly couple to the e' HOMO in 1bn. In addition to this mode, the E' modes of 520 and 1399 cm^{-1} and the A_1' mode of 949 cm^{-1} can also couple to the e' HOMO in 1bn. But it should be noted that the E' mode of 1399 cm^{-1} much less strongly couples to the e' HOMO than the E' mode of 1492 cm^{-1} even though both vibrational modes are the B–N stretching modes. This can be understood as follows. The reduced mass for the E' mode of 1399 cm^{-1} is 1.89, while that for the E' mode of 1492 cm^{-1} is 4.93, and thus the displacements of B and N atoms are larger in the E' mode of 1492 cm^{-1} than in the E' mode of 1399 cm^{-1}. The e' HOMO is, on the other hand, localized on B and N atoms. It is rational that a frequency mode in which the displacements of B and N atoms are important, more strongly couples to the e' HOMO localized on B and N atoms. This is the reason why the E' mode of 1399 cm^{-1} much less strongly couples to the e' HOMO than the E' mode of 1492 cm^{-1} in 1bn. It should be noted that the B–H stretching E' mode of 2636 cm^{-1} somewhat couples to the e' HOMO in 1bn. Considering that the electron density on boron and hydrogen atoms in the e' HOMO is not high, such electron–phonon coupling constant for the E' mode of 2636 cm^{-1} is rather large.

Let us next look into the electron–phonon coupling in the monocations of 2bn and 3bn. We can see from Figure 8 that the B–N stretching A_1 mode of 1486 cm^{-1} can strongly couple to the HOMO in 2bn. Furthermore, the B–N stretching A_1 modes of 1480 and 1528 cm^{-1} strongly couple to the HOMO in 3bn. It should be noted that the electron–phonon coupling constants for the low frequency modes decrease with an increase in molecular size from the monocations of 1bn to 3bn more rapidly than those for the B–N stretching modes around 1500 cm^{-1}.

Let us next compare the calculated results for the monocations of B, N-substituted polyacenes with those for the monocations of polyacenes. We can see from Figure 8 that as in the monocation of 1bn, both the low frequency modes and the C–C stretching modes around 1500 cm^{-1} afford large electron–phonon coupling constants in the monocation of 1a. But in the monocations of large size of polyacenes such as 2a and 3a, only the C–C stretching modes around 1500 cm^{-1} afford large electron–phonon coupling constants, and the electron–phonon coupling constants for the C–C stretching modes around 1500 cm^{-1} decrease with an increase in molecular size from the monocations of 2a to 6a.

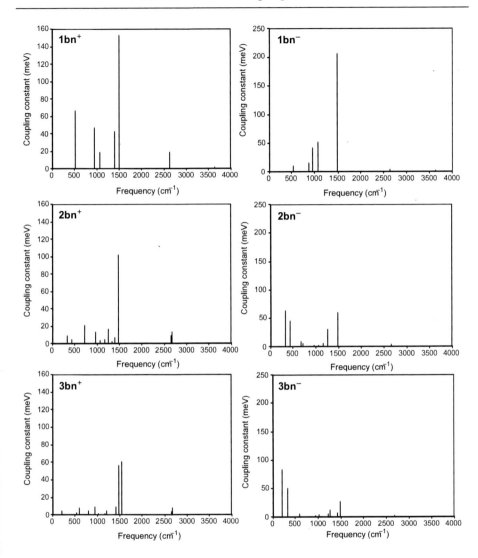

Figure 8. Electron–phonon coupling constants for the monocations and monoanions of B, N-substituted polyacenes.

2. Monoanions

We can see from Figure 8 that the B–N stretching E' mode of 1492 cm^{-1} can strongly couple to the e' LUMO in 1bn. This can be understood in view of the phase patterns of the LUMO in 1bn. When 1bn is distorted by the E' mode of

1492 cm^{-1} toward the same direction as shown in Figure 3, the bonding (antibonding) interactions between two neighboring B and N atoms in the LUMO (a) become weaker (stronger), therefore, the LUMO (a) is significantly destabilized in energy. In a similar way, the antibonding interactions between two neighboring B and N atoms in the LUMO (b) become weaker by such a distortion, therefore, the LUMO (b) is significantly stabilized in energy. That is the reason why the E' mode of 1492 cm^{-1} can strongly couple to the e' LUMO in 1bn. In addition to this mode, the E' mode of 1073 cm^{-1} and the A_1' mode of 949 cm^{-1} can also couple to the e' LUMO in 1bn. But it should be noted that the B–N stretching E' mode of 1399 cm^{-1} hardly couples to the e' LUMO in 1bn. This can be understood as follows. When 1bn is distorted by the E' mode of 1399 cm^{-1} toward the same direction as shown in Figure 3, the bonding (antibonding) interaction between N(1) and B(2) becomes stronger, while the bonding (antibonding) interaction between B(3) and N(3) becomes weaker in the LUMO (a) (LUMO (b)). Since such changes in the strengths of orbital interactions are compensated by each other, the energy levels of the LUMO (a) and LUMO (b) are not significantly changed by such a distortion. That is the reason why the B–N stretching E' mode of 1399 cm^{-1} hardly couples to the e' LUMO in 1bn. Let us next look into the electron–phonon coupling in the monoanions of 2bn and 3bn. We can see from Figure 8 that the lowest frequency A_1 modes of 327 and 447 cm^{-1} and the B–N stretching mode of 1486 cm^{-1} can strongly couple to the a_2 LUMO in 2bn. Furthermore, the lowest frequency A_1 modes of 212 and 345 cm^{-1} and the B–N stretching A_1 mode of 1480 cm^{-1} can strongly couple to the b_1 LUMO in 3bn. It should be noted that the B–N stretching modes around 1500 cm^{-1} and the low frequency modes, less and more, respectively, strongly couple to the LUMO with an increase in molecular size from 1bn to 3bn.

Let us next compare the calculated results for the monocations with those for the monoanions [74] in B, N-substituted polyacenes. As described above, the HOMO and LUMO in B, N-substituted polyacene-series are rather localized on N and B atoms, respectively, due to the electronegativity perturbation on polyacene-series. Therefore, significant vibronic interactions character differences between the monoanions and cations in B, N-substituted polyacene-series can be expected. As can be seen from Figure 8, both the low frequency modes and the B–N stretching modes around 1500 cm^{-1} afford large electron–phonon coupling constants in the monoanions and cations in B, N-substituted polyacenes. The low frequency modes play more important role in the electron–phonon interactions in the monocation than in the monoanion in 1bn. But the electron–phonon coupling constants for the low frequency modes decrease more rapidly with an increase in molecular size than those for the B–N stretching modes around 1500 cm^{-1} in the monocations of B, N-substituted polyacene-series. Therefore, the B–N stretching modes would play more important role in the electron–phonon interactions than the low frequency modes even in the large size of the monocations of B, N-substituted polyacenes. But the B–N stretching modes around 1500 cm^{-1} and the low frequency modes, less and more, respectively, strongly couple to the LUMO with an increase in molecular size from 1bn to 3bn. Therefore, the higher frequency modes would play an essential role in the electron–phonon interactions in the large size of the monocations, while the lower frequency modes play such a role in the large size of the monoanions in B, N-substituted polyacenes. This difference may come from the phase patterns difference between the HOMO and LUMO due to electronegativity perturbation on polyacene-series.

We can therefore conclude in this section that the lower frequency modes play an important role in the electron–phonon interactions in large size of the negatively charged polyacene-series and B, N-substituted polyacene-series, while the higher frequency modes such as B–N stretching modes around 1500 cm^{-1} play such a role in the positively charged polyacene-series and B, N-substituted polyacene-series.

D. Polycyanodienes

1. Monocations

The calculated electron–phonon coupling constants in the monocations of 2cn, 3cn, and 4cn are shown in Figure 9. Let us next look into the electron–phonon interactions between the a_g HOMO and the A_g modes in 2cn. We can see from Figure 9 that the A_g mode of 1597 cm^{-1} the most strongly couples to the a_g HOMO in 2cn. This can be understood in view of the phase patterns of the HOMO in 2cn. When 2cn is distorted along the A_g mode of 1597 cm^{-1} toward the same direction as shown in Figure 3, the bonding interactions between C(3) and C(4) atoms in the HOMO become weaker, and the antibonding interactions between N(2) and C(3) atoms become stronger. Furthermore, the characteristics of the bonding and antibonding interactions between two neighboring carbon and hydrogen atoms in the HOMO become less and more, respectively, significant.

Therefore, the a_g HOMO is significantly destabilized in energy by such a distortion. This is the reason why the A_g mode of 1597 cm^{-1} the most strongly couples to the a_g HOMO in 2cn. In addition to this, the lowest frequency A_g mode of 555 cm^{-1} affords large electron–phonon coupling constant in the monocation of 2cn. When 2cn is distorted along the A_g mode of 555 cm^{-1}, the characteristics of the antibonding interactions between N(2) and C(3) atoms become more significant, and the bonding interactions between two neighboring carbon and hydrogen atoms in the HOMO become weaker. Therefore, the a_g HOMO is significantly destabilized in energy by such a distortion. This is the reason why the A_g mode of 555 cm^{-1} affords large electron–phonon coupling constant in the monocation of 2cn. Similar discussions can be made in 3cn and 4cn; the C–C and C–N stretching modes around 1500 cm^{-1} and the lowest frequency modes strongly couple to the HOMO.

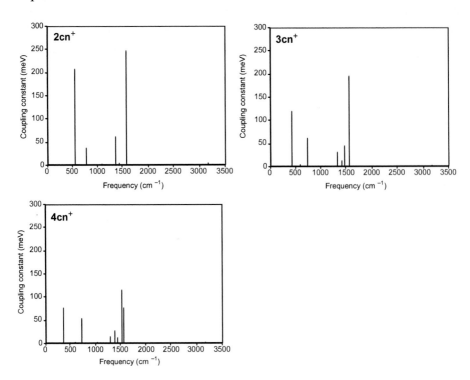

Figure 9. Electron–phonon coupling constants for the monocations of polycyanodienes.

2. Monoanions

The calculated electron–phonon coupling constants in the monoanions of 1cn, 2cn, 3cn, and 4cn are shown in Figure 10. Let us first look into the electron–phonon interactions between the A_g modes and the b_{3u} LUMO in 1cn. We can see from Figure 10 that the A_g mode of 1713 cm^{-1} can the most strongly couple to the

b_{3u} LUMO in 1cn. When 1cn is distorted along the A_g mode of 1713 cm^{-1} toward the same direction as shown in Figure 3, the bonding interactions between two neighboring carbon atoms in the b_{3u} LUMO become stronger, and the antibonding interactions between two neighboring carbon and nitrogen atoms become weaker, and thus the b_{3u} LUMO is significantly stabilized in energy. This is the reason why the A_g mode of 1713 cm^{-1} can the most strongly couple to the b_{3u} LUMO in 1cn. The A_g mode of 1035 cm^{-1} also strongly couples to the b_{3u} LUMO in 1cn. When 1cn is distorted along the A_g mode of 1035 cm^{-1}, the antibonding interactions between two neighboring carbon and nitrogen atoms in the b_{3u} LUMO become significantly weaker, and thus the b_{3u} LUMO is significantly stabilized in energy by such a distortion. On the other hand, the bonding interactions between two neighboring carbon atoms become also slightly weaker by such a distortion, the b_{3u} LUMO would be slightly destabilized in energy. Such stabilized and destabilized effects are compensated by each other. Therefore, the A_g mode of 1035 cm^{-1} affords slightly smaller electron–phonon coupling constant than the A_g mode of 1713 cm^{-1} in the monoanion of 1cn. Furthermore, the A_g mode of 623 cm^{-1} also somewhat strongly couples to the b_{3u} LUMO in 1cn. The reduced masses for the A_g modes of 623, 1035 and 1713 cm^{-1} are large, and are 8.80, 7.53 and 5.97, respectively, in 1cn. Therefore, the displacements of carbon and nitrogen are large in these modes. It is rational that the A_g modes of 623, 1035, and 1713 cm^{-1}, in which the displacements of carbon and nitrogen atoms are larger, can strongly couple to the b_{3u} LUMO localized on carbon and nitrogen atoms. On the other hand, the reduced masses for the A_g modes of 1252 and 3200 cm^{-1} are 1.11 and 1.10, respectively. It is rational that the A_g modes of 1252 and 3200 cm^{-1} in which the displacements of carbon and nitrogen atoms are small cannot strongly couple to the b_{3u} LUMO localized on carbon and nitrogen atoms. This is the reason why the A_g modes of 623, 1035, and 1713 cm^{-1} can more strongly couple to the b_{3u} LUMO than the A_g modes of 1252 and 3200 cm^{-1} in 1cn.

Let us look into the electron–phonon interactions between the A_g modes and the b_{1g} LUMO in 2cn. We can see from Figure 10 that the A_g mode of 1597 cm^{-1} strongly couples to the b_{1g} LUMO in 2cn. This can be understood as follows. When 2cn is distorted along the A_g mode of 1597 cm^{-1}, the bonding interactions between two neighboring carbon atoms in the b_{1g} LUMO become significantly weaker, and the antibonding interactions between two neighboring carbon and nitrogen atoms become significantly stronger, and thus the b_{1g} LUMO is significantly destabilized in energy by such a distortion in 2cn. This is the reason why the A_g mode of 1597 cm^{-1} can strongly couple to the b_{1g} LUMO in 2cn. The A_g mode of 1436 cm^{-1} can also strongly couple to the b_{1g} LUMO in

2cn. When 2cn is distorted along the A_g mode of 1436 cm^{-1}, the antibonding interactions between two neighboring carbon and nitrogen atoms in the b_{1g} LUMO in 2cn become stronger, and thus the b_{1g} LUMO is significantly destabilized in energy by such a distortion in 2cn.

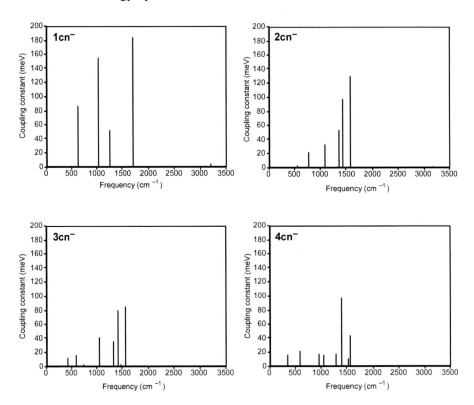

Figure 10. Electron–phonon coupling constants for the monoanions of polycyanodienes.

This is the reason why the A_g mode of 1436 cm^{-1} strongly couples to the b_{1g} LUMO in 2cn. Apart from the C–H stretching A_g mode of 3187 cm^{-1}, the electron–phonon coupling constant increases with an increase in frequency in 2cn. In the low frequency modes, which have similar characteristics to those of acoustic mode of phonon in solids, all atoms move toward the similar direction. Therefore, the orbital interactions between two neighboring atoms do not significantly change when 2cn is distorted along the low frequency A_g modes. This is the reason why the electron–phonon coupling decreases with a decrease in frequency in 2cn. The C–H stretching mode of 3187 cm^{-1} hardly couples to the b_{1g} LUMO in 2cn. The reduced mass for the A_g mode of 3187 cm^{-1} is 1.10, and in which the displacements of carbon and nitrogen atoms are very small. The b_{1g} LUMO is localized on carbon and nitrogen atoms in 2cn. It is rational that the A_g mode of 3187 cm^{-1}, in which the displacements of carbon and nitrogen atoms are very small, hardly couples to the b_{1g} LUMO localized on carbon and nitrogen atoms in 2cn. Similar discussions can be made in the monoanions of 3cn and 4cn; the C–C stretching modes around 1500 cm^{-1} can strongly couple to the LUMO.

E. Summary

In this chapter, we investigated the electron–phonon interactions in the monocations and monoanions of polyacenes, polyfluoroacenes, B, N-substituted polyacenes, and polycyanodienes. The C–C stretching modes around 1500 cm^{-1} strongly couple to the HOMO, and the lowest frequency modes and the C–C stretching modes around 1500 cm^{-1} strongly couple to the LUMO in polyacenes. The C–C stretching modes around 1500 cm^{-1} strongly couple to the HOMO and LUMO in polyfluoroacenes. The B–N stretching modes around 1500 cm^{-1} strongly couple to the HOMO and LUMO in B, N-substituted polyacenes. The C–C and C–N stretching modes around 1500 cm^{-1} strongly couple to the HOMO and LUMO in polycyanodienes.

Chapter VI

TOTAL ELECTRON–PHONON COUPLING CONSTANTS

A. Polyacenes and Polyfluoroacenes

1. Monocations

Let us next discuss the total electron–phonon coupling constants in the monocations (l_{HOMO}) of polyfluoroacenes, and compare the calculated results for polyfluoroacenes with those for polyacenes. The l_{HOMO} values for *1a* and *1fa* are defined as

$$l_{HOMO} = \sum_{m=1}^{6} l_{HOMO}(\omega_m) = \sum_{m=1}^{4} g^2_{HOMO}(\omega_m)\hbar\omega_m + \frac{1}{2}\sum_{m=5}^{6} g^2_{HOMO}(\omega_m)\hbar\omega_m \tag{55}$$

and those for polyacenes and polyfluoroacenes (*2fa–5fa* and *2fa–5fa*) are defined as

$$l_{HOMO} = \sum_{m} l_{HOMO}(\omega_m) = \sum_{m} g^2_{HOMO}(\omega_m)\hbar\omega_m \tag{56}$$

The calculated total electron–phonon coupling constants (l_{LUMO} and l_{HOMO}) in the monoanions and cations of polyacenes, polyfluoroacenes, B, N-substituted polyacenes, and polycyanodienes are shown in Figure 11. The l_{HOMO} values are estimated to be 0.418, 0.399, 0.301, 0.255, and 0.222 eV for 1fa, 2fa, 3fa, 4fa, and 5fa, respectively, and those are estimated to be 0.244, 0.173, 0.130, 0.107, and 0.094 eV for 1a, 2a, 3a, 4a, and 5a, respectively. Therefore, the l_{HOMO} values decrease with an increase in molecular size in both polyacenes and polyfluoroacenes. This can be understood as follows. The electron density per atom in the HOMO becomes lower with an increase in molecular size, and the orbital interactions between two adjacent atoms become weaker with an increase in number of atoms in polyacenes and polyfluoroacenes. Therefore, strengths of the orbital interactions between two adjacent atoms less significantly change with an increase in number of atoms when these monocations are distorted along the C–C stretching modes around 1500 cm^{-1} playing an essential role in the electron–phonon interactions. This is the reason why the l_{HOMO} value decreases with an increase in number of atoms. Therefore, in general, we can expect that a monocation, in which number of carriers per atom is larger, affords larger l_{HOMO} value.

The l_{HOMO} values for polyfluoroacenes are much larger than those for polyacenes. This can be understood as follows. For example, let us compare the calculated results for 2fa with those for 2a. The reduced masses of the C–C stretching A_g modes of 1445 and 1610 cm^{-1} in 2a, which afford large electron–phonon coupling constants in the monocation of 2a, are 9.68 and 7.19, respectively, and thus the displacements of carbon atoms are not very large in these vibrational modes. On the other hand, the reduced masses of the C–C stretching modes of 1552 and 1666 cm^{-1} in 2fa, which afford large electron–phonon coupling constants in the monocation of 2fa, are 12.45 and 12.10, respectively, and thus the displacements of carbon atoms are very large in these vibrational modes. It is rational that the C–C stretching modes around 1500 cm^{-1}, in which the displacements of carbon atoms are larger in 2fa, can more strongly couple to the HOMO, in which the electron density on carbon atoms are very high, than the C–C stretching modes around 1500 cm^{-1}, in which the displacements of carbon atoms are smaller in 2a. Furthermore, the low frequency mode of 514 cm^{-1} affords much larger electron–phonon coupling constant in the monocation of 2fa than the low frequency modes around 500 cm^{-1} in the monocation of 2a. Similar discussion can be made in 3fa, 4fa, and 5fa.

Therefore, we can conclude that the larger values of the electron–phonon coupling constants for the C–C stretching modes around 1500 cm^{-1} in the monocations of polyfluoroacenes than those in the monocations of polyacenes due to the larger displacements of carbon atoms in the C–C stretching modes in polyfluoroacenes than those in polyacenes, and the larger values of the electron–

phonon coupling constants for the low frequency modes around 500 cm^{-1} in the monocations of polyfluoroacenes than those in the monocations of polyacenes, are the main reason why the l_{HOMO} values for polyfluoroacenes are much larger than those for polyacenes.

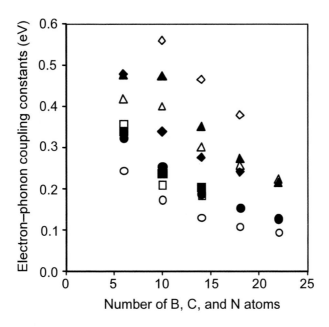

Figure 11. Total electron–phonon coupling constants for the monocations and monoanions of polyacenes, polyfluoroacenes, B, N-substituted polyacenes, and polycyanodienes. The opened circles, triangles, squares, and diamonds represent the l_{HOMO} values for polyacenes, polyfluoroacenes, B, N-substituted polyacenes, and polycyanodienes, respectively, and the closed circles, triangles, squares, and diamonds represent the l_{LUMO} values for polyacenes, polyfluoroacenes, B, N-substituted polyacenes, and polycyanodienes, respectively.

2. Monoanions

Let us next discuss the total electron–phonon coupling constants in the monoanions (l_{LUMO}) of polyfluoroacenes, and compare the calculated results for polyfluoroacenes with those for polyacenes. The l_{LUMO} value for **1fa** is defined as

$$l_{\text{LUMO}} = \sum_{m=1}^{6} l_{\text{LUMO}}(\omega_m) = \sum_{m=1}^{4} g^2_{\text{LUMO}}(\omega_m)\widetilde{\omega}_m + \frac{1}{2}\sum_{m=5}^{6} g^2_{\text{LUMO}}(\omega_m)\widetilde{\omega}_m$$

(57)

and that for polyfluoroacenes (*2fa–5fa*) is defined as

$$l_{\text{LUMO}} = \sum_{m} l_{\text{LUMO}}(\omega_m) = \sum_{m} g^2_{\text{LUMO}}(\omega_m)\widetilde{\omega}_m$$

(58)

The l_{LUMO} values are estimated to be 0.475, 0.473, 0.350, 0.273, and 0.215 eV for 1fa, 2fa, 3fa, 4fa, and 5fa, respectively, while those are estimated to be 0.322, 0.255, 0.186, 0.154, and 0.127 eV for 1a, 2a, 3a, 4a, and 5a, respectively. Therefore, the l_{LUMO} values decrease with an increase in molecular size in both polyacenes and polyfluoroacenes.

The l_{LUMO} values for polyfluoroacenes are much larger than those for polyacenes. This can be understood as follows. As described above, the LUMOs of polyfluoroacenes are rather localized on carbon atoms, and those of polyacenes are completely localized on carbon atoms. And the displacements of carbon atoms in the C–C stretching modes around 1500 cm^{-1} in polyfluoroacenes become larger than those in polyacenes as a consequence of H–F substitution in polyacenes. That is, by substituting hydrogen atoms by heavier atoms which have the highest electronegativity in every element, fluorine atoms, the displacements of carbon atoms become significantly larger in the C–C stretching modes, while the electronic structure in the LUMO hardly changes. Therefore, the C–C stretching modes around 1500 cm^{-1} in polyfluoroacenes more strongly couple to the LUMO localized on carbon atoms than those in polyacenes. This is the main reason why the l_{LUMO} values for polyfluoroacenes are much larger than those for polyacenes. The l_{LUMO} value for polyfluoroacenes decreases with an increase in molecular size more rapidly than that for polyacenes, and the l_{LUMO} values difference between polyacenes and polyfluoroacenes decreases with an increase in molecular size. This is because the ratio of the number of fluorine atoms to that of carbon atoms becomes smaller from 1fa to 5fa, and the effects of H–F substitution become less important in this order.

Let us next compare the calculated results for the monocations with those for the monoanions in polyfluoroacenes. The l_{LUMO} values are larger than the l_{HOMO} values in 1fa, 2fa, 3fa, and 4fa. This can be understood as follows. The C–C stretching modes around 1500 cm^{-1}, in which the displacements of carbon atoms are very large, can more strongly couples to the LUMO rather localized on carbon atoms than to the delocalized HOMO in polyfluoroacenes. This is the reason why the electron–phonon coupling constants for the C–C stretching A_g modes around 1400 cm^{-1} in the monoanions are larger than those for the C–C stretching A_g modes around 1600 cm^{-1} in the monocations in polyfluoroacenes, and the reason why the l_{LUMO} values are larger than the l_{HOMO} values in 1fa, 2fa, 3fa, and 4fa. But it should be noted that the l_{LUMO} value decreases with an increase in molecular size more rapidly than the l_{HOMO} value in polyfluoroacenes, and the l_{LUMO} value is slightly smaller than the l_{HOMO} value in 5fa.

The l_{HOMO} and l_{LUMO} values for polyfluoroacene with D_{2h} geometry would converge and be estimated by extrapolation in Figure 11 to be 0.074 and 0.009 eV, respectively, assuming that the l_{HOMO} and l_{LUMO} values are approximately inversely proportional to the number of carbon atoms in each series.

The l_{HOMO} value is estimated to be much larger than the l_{LUMO} value in polyfluoroacene ($N \to \infty$), even though the l_{HOMO} value is smaller than the l_{LUMO} value in small size of polyfluoroacenes such as 2fa, 3fa, and 4fa. That is, the l_{HOMO} values decrease with an increase in molecular size less rapidly than the l_{LUMO} values in polyfluoroacenes. This is because the electron–phonon coupling constants originating from the C–C stretching modes around 1500 cm^{-1} decrease with an increase in molecular size more rapidly in the monoanions than in the monocations in polyfluoroacenes. The strengths of the orbital interactions between two neighboring carbon atoms in the HOMO and LUMO significantly change when polyfluoroacenes are distorted along the C–C stretching modes around 1500 cm^{-1}, and thus the C–C stretching modes around 1500 cm^{-1} can strongly couple to the HOMO and LUMO in polyfluoroacenes. Strengths of such orbital interactions in the HOMO become weaker with an increase in molecular size less rapidly than those in the LUMO in polyfluoroacenes because the electron density on carbon atoms in the LUMO rather localized on carbon atoms decrease with an increase in molecular size more rapidly than those in the delocalized HOMO in polyfluoroacenes. This is the reason why the electron–phonon coupling constants originating from the C–C stretching modes around 1500 cm^{-1} decrease with an increase in molecular size more rapidly in the monoanions than in the monocations in polyfluoroacenes, and the reason why the l_{HOMO} values decrease with an increase in molecular size less rapidly than the l_{LUMO} values in polyfluoroacenes.

The total electron–phonon coupling constants were estimated to be 0.322, 0.254, 0.186, 0.154, 0.127, and 0.106 eV (0.244, 0.173, 0.130, 0.107, 0.094, and

0.079 eV) in the monoanions (monocations) of 1a, 2a, 3a, 4a, 5a, and 6a, respectively. Therefore, the electron–phonon coupling constants decrease with an increase in molecular size. Figures 4 and 5 demonstrate that the electron–phonon coupling originating from the C–C stretching modes of 1400–1600 cm^{-1} becomes weaker as the molecular size becomes larger from 1a to 6a. Thus, the C–C stretching modes should play an important role in the electron–phonon coupling in the monoanions and cations of 1a and polyacenes. The total electron–phonon coupling constants in the monocations of polyacenes (l_{HOMO}) are smaller than those in the monoanions (l_{LUMO}). This is because the lowest frequency mode, which plays an important role in the electron–phonon coupling in the monoanions of polyacenes, does not play a role in the electron–phonon coupling in the monocations.

B. B, N-Substituted Polyacenes

The l_{HOMO} for *1bn* is defined as

$$l_{HOMO} = \sum_{m=1}^{11} l_{HOMO}(\omega_m) = \sum_{m=1}^{7} g_{HOMO}^2(\omega_m)\hbar\omega_m + \frac{1}{2}\sum_{m=8}^{11} g_{HOMO}^2(\omega_m)\hbar\omega_m \tag{59}$$

That for 2bn is defined as

$$l_{HOMO} = \sum_{m=1}^{17} l_{HOMO}(\omega_m) = \sum_{m=1}^{17} g_{HOMO}^2(\omega_m)\hbar\omega_m \tag{60}$$

and that for 3bn is defined as

$$l_{HOMO} = \sum_{m=1}^{23} l_{HOMO}(\omega_m) = \sum_{m=1}^{23} g_{HOMO}^2(\omega_m)\hbar\omega_m \tag{61}$$

The l_{LUMO} for *1bn* is defined as

$$l_{LUMO} = \sum_{m=1}^{11} l_{LUMO}(\omega_m) = \sum_{m=1}^{7} g_{LUMO}^2(\omega_m)\hbar\omega_m + \frac{1}{2}\sum_{m=8}^{11} g_{LUMO}^2(\omega_m)\hbar\omega_m \tag{62}$$

That for 2bn is defined as

$$l_{\text{LUMO}} = \sum_{m=1}^{17} l_{\text{LUMO}}(\omega_m) = \sum_{m=1}^{17} g_{\text{LUMO}}^2(\omega_m) \hbar\omega_m \tag{63}$$

and that for 3bn is defined as

$$l_{\text{LUMO}} = \sum_{m=1}^{23} l_{\text{LUMO}}(\omega_m) = \sum_{m=1}^{23} g_{\text{LUMO}}^2(\omega_m) \hbar\omega_m \tag{64}$$

The l_{HOMO} (l_{LUMO}) values are estimated to be 0.357, 0.209, and 0.182 eV (0.340, 0.237, and 0.203 eV) for 1bn, 2bn, and 3bn, respectively, while those are estimated to be 0.244, 0.173, and 0.130 eV (0.322, 0.254, and 0.186 eV) for 1a, 2a, and 3a, respectively. Therefore, both the l_{HOMO} and l_{LUMO} values decrease with an increase in molecular size in B, N-substituted polyacene-series as well as in polyacene-series.

The l_{LUMO} values are much larger than the l_{HOMO} values in polyacene-series, while the l_{HOMO} values are similar to the l_{LUMO} values in B, N-substituted polyacene-series. This can be understood as follows. The low frequency modes as well as the C–C stretching modes around 1500 cm^{-1} afford electron–phonon coupling constants in the monoanions of polyacenes, while only the C–C stretching modes around 1500 cm^{-1} afford large electron–phonon coupling constants in the monocations of polyacenes except for 1a. Furthermore, the C–C stretching modes around 1500 cm^{-1} and the low frequency modes, less and more, respectively, strongly couple to the LUMO, while the C–C stretching modes around 1500 cm^{-1} less strongly couple to the HOMO with an increase in molecular size in polyacene-series. The electron–phonon coupling originating from the low frequency modes in the monoanions is the main reason why the l_{LUMO} values are much larger than the l_{HOMO} values in polyacene-series. Let us next look into B, N-substituted polyacene-series. The B–N stretching modes around 1500 cm^{-1} as well as the low frequency modes afford large electron–phonon coupling constants both in the monoanions and cations of B, N-substituted polyacene-series. The B–N stretching modes around 1500 cm^{-1} and the low frequency modes, less and more, respectively, strongly couple to the LUMO with an increase in molecular size, while the low frequency modes much less strongly couple to the HOMO than the B–N stretching modes around 1500 cm^{-1} with an increase in molecular size in B, N-substituted polyacene-series. Even

Takashi Kato

though the low and high frequency modes afford large electron–phonon coupling constants in the monoanions and cations in B, N-substituted polyacene-series, respectively, there are not significant differences between the l_{LUMO} and l_{HOMO} values. This is the reason why the l_{LUMO} values are similar to the l_{HOMO} values in B, N-substituted polyacene-series.

We can see from Figure 11 that the l_{HOMO} values for B, N-substituted polyacenes are larger than those for polyacenes, and can expect that those for the large size of B, N-substituted polyacene-series are larger than those for the large size of polyacene-series. The l_{HOMO} values for B, N-substituted polyacene ($N \to \infty$) and polyacene ($N \to \infty$) are estimated to be 0.096 and 0.036 eV, respectively, assuming that the l_{HOMO} is approximately inversely proportional to the number of C, B, and N atoms in each series.[65] This can be understood as follows. For example, let us compare the HOMO of 3bn with that of 3a. We can see from Figure 3 that the HOMO of 3a is rather localized on C(1), C(3), C(5), C(8), C(10) and C(12) atoms which are located at the edge of the carbon framework, and the electron density on other carbon atoms is very low. Therefore, the HOMO of 3a has a non-bonding character and the orbital interactions between two neighboring carbon atoms are not so large. Such a non-bonding character in the HOMO becomes more significant with an increase in molecular size in polyacene-series. This is the reason why the l_{HOMO} values decrease with an increase in molecular size in polyacene-series. Let us next look into 3bn. Because of the electronegativity perturbation on 3a, the HOMO of 3bn is stabilized in energy with respect to the HOMO of 3a, and the electron density on N(7), N(9), N(11), and N(13) atoms in 3bn is higher than that on C(7), C(9), C(11), and C(13) atoms in 3a, respectively. Therefore, orbital interactions between N(7) and B(8), between B(8) and N(9), between N(9) and B(10), between B(10) and N(11), between N(11) and B(12), and between B(12) and N(13) atoms in the HOMO of 3bn, are stronger than those between two neighboring carbon atoms in the HOMO of 3a. This is the reason why the l_{HOMO} value for 3bn is larger than that for 3a. In a similar way, the large size of B, N-substituted polyacene-series would have larger l_{HOMO} values than the large size of polyacene-series. We can therefore conclude that because of the strong orbital interactions between B and N atoms in the HOMO in B, N-substituted polyacene-series as a consequence of the electronegativity perturbation on polyacene-series, the l_{HOMO} values for B, N-substituted polyacene-series would become larger than those for polyacene-series.

C. Polycyanodienes

1. Monocations

Let us next discuss the total electron–phonon coupling constants in the monocations (l_{HOMO}) of polycyanodienes, and compare the calculated results for polycyanodienes with those for polyacenes. The l_{HOMO} values for polycyanodienes are defined as

$$l_{HOMO} = \sum_m l_{HOMO}(\omega_m) = \sum_m g^2_{HOMO}(\omega_m)\hbar\omega_m \qquad (65)$$

The l_{HOMO} values are estimated to be 0.558, 0.464, and 0.378 eV for 2cn, 3cn, and 4cn, respectively. Those were estimated to be 0.173, 0.130, and 0.107 eV for 2a, 3a, and 4a, respectively. Therefore, the l_{HOMO} values decrease with an increase in molecular size in the monocations of polycyanodienes as well as in the monocations of polyacenes. The l_{HOMO} values for polycyanodienes are much larger than those for polyacenes. This can be understood as follows. The HOMO of polyacenes is localized on edge part of carbon atoms, and thus the nonbonding characteristics are significant in the HOMO of polyacenes. Therefore, the energy levels of the HOMO in polyacenes do not significantly change even if polyacenes are distorted along the C–C stretching modes around 1500 cm^{-1} playing an essential role in the electron–phonon interactions. As in case of the HOMO of polyacenes, the HOMO of polycyanodienes are also somewhat localized on edge part of nitrogen atoms. On the other hand, the σ orbital interactions as well as the π orbital interactions are significant in the HOMO of polycyanodienes. Therefore, orbital interactions between two neighboring atoms are much stronger in the HOMO of polycyanodienes than in the HOMO of polyacenes. Furthermore, while the HOMO of polyacenes are completely localized on carbon atoms, the HOMO of polycyanodienes are delocalized, and the electron density on hydrogen atoms as well as that on carbon and nitrogen atoms in the HOMO of polycyanodienes is high. Therefore, the strengths of the orbital interactions between two neighboring carbon and hydrogen atoms are significantly changed by distortions by vibronic active modes. This is the reason why the l_{HOMO} values for σ-conjugated polycyanodienes are much larger than those for π-conjugated polyacenes. In summary, the strong σ orbital interactions between two neighboring carbon atoms and two neighboring carbon and nitrogen atoms and the orbital interactions between two neighboring carbon and hydrogen atoms in the HOMO of polycyanodienes are the main reason why the l_{HOMO} values for σ-conjugated polycyanodienes are much larger than those for π-conjugated polyacenes.

Let us next estimate the l_{HOMO} values for polycyanodienes ($N \to \infty$) with D_{2h} geometry. Assuming that the l_{HOMO} values for polycyanodienes with D_{2h} geometry are approximately inversely proportional to the number of carbon and nitrogen atoms in each series, as suggested in previous research, [21] the l_{HOMO} value for polycyanodiene ($N \to \infty$) is estimated to be 0.167 eV. The l_{HOMO} value for polyacene ($N \to \infty$) with D_{2h} geometry is estimated to be 0.027 eV. Therefore, the l_{HOMO} values for polycyanodienes ($N \to \infty$) are estimated to be much larger than those for polyacenes ($N \to \infty$). Strong σ orbital interactions between two neighboring atoms and the orbital interactions between two neighboring carbon and hydrogen atoms in the HOMO of polycyanodienes are the main reason why the l_{HOMO} values for σ-conjugated polycyanodienes are much larger than those for π-conjugated polyacenes.

2. Monoanions

Let us next discuss the total electron–phonon coupling constants in the monoanions (l_{LUMO}) of polycyanodienes, and compare the calculated results for polycyanodienes with those for polyacenes. The l_{LUMO} values for polycyanodienes are defined as

$$l_{LUMO} = \sum_m l_{LUMO}(\omega_m) = \sum_m g^2_{LUMO}(\omega_m)\hbar\omega_m ,$$

(66)

The l_{LUMO} values are estimated to be 0.478, 0.338, 0.275, and 0.240 eV for 1cn, 2cn, 3cn, and 4cn, respectively, while those are estimated to be 0.322, 0.255, 0.186, and 0.154 eV for 1a, 2a, 3a, and 4a, respectively, and those are estimated to be 0.340, 0.237, and 0.203 eV for 1bn, 2bn, and 3bn, respectively. Therefore, the l_{LUMO} values decrease with an increase in molecular size in polyacenes, B, N-substituted polyacenes, and polycyanodienes.

The l_{LUMO} values for polycyanodienes are much larger than those for polyacenes. This can be understood in view of the phase patterns of the LUMO in polyacenes and polycyanodienes. The LUMOs of polyacenes are rather localized on carbon atoms located at the edge part of carbon framework. Therefore, non-bonding characteristic is significant in the LUMOs in polyacenes. On the other hand, the LUMO of polycyanodienes is delocalized, and the electron density on carbon atoms as well as on nitrogen atoms located at the edge part of CN framework is high in the LUMO in polycyanodienes. Therefore, the orbital interactions between two neighboring atoms in the LUMO in polycyanodienes are stronger than those in the LUMO in polyacenes. This is the reason why the l_{LUMO} values for polycyanodienes are much larger than those for polyacenes. Let us next discuss why the LUMOs of polyacenes are rather localized on carbon atoms located at the edge part of carbon framework, while those of polycyanodienes are delocalized, and the electron density on carbon atoms as well as on nitrogen atoms is high in the LUMO in polycyanodienes. In general, due to electronegativity perturbation, [30] the π bonding orbitals are weighted more heavily on the atoms which are more electronegative, and the π antibonding orbitals weighted more heavily on the atoms which is less electronegative. As described above, the LUMO are rather localized on carbon atoms which are located at the edge part of carbon framework in polyacenes. In polycyanodienes, carbon atoms with high electron density in the LUMO, located at the edge part of carbon framework of polyacenes, are substituted by nitrogen atoms with the higher electronegativity. Therefore, the electron density on nitrogen atoms in the LUMO in polycyanodienes is lower than that on carbon atoms located at the edge part of carbon framework in the LUMO of polyacenes, and the electron density on carbon atoms which are not located at the edge part of CN framework in the LUMO in polycyanodienes is higher than that on carbon atoms which are not located at the edge part of carbon framework in the LUMO in polyacenes. This is the reason why the LUMOs of polyacenes are rather localized on carbon atoms located at the edge part of carbon framework, while those of polycyanodienes are delocalized, and the electron density on carbon atoms as well as on nitrogen atoms is high in polycyanodienes.

Let us next estimate the l_{LUMO} values for polycyanodienes ($N \to \infty$) with D_{2h} geometry. Assuming that the l_{LUMO} values for polycyanodienes with D_{2h} geometry are approximately inversely proportional to the number of carbon and nitrogen atoms in each series, as suggested before, the l_{LUMO} value for polycyanodiene ($N \to \infty$) is estimated to be 0.122 eV. The l_{LUMO} value for polyacene ($N \to \infty$) with D_{2h} geometry was estimated to be 0.019 eV.

In summary, the l_{LUMO} values for polycyanodienes are estimated to be much larger than those for polyacenes. Therefore, the orbital patterns difference between the LUMO localized on carbon atoms located at the edge part of carbon framework in polyacenes and the delocalized LUMO in polycyanodienes due to electronegativity perturbation, is the main reason why the l_{LUMO} values for polycyanodienes are much larger than those for polyacenes with D_{2h} geometry.

D. Summary

In this chapter, we estimated the total electron–phonon coupling constants in the monocations and monoanions of polyacenes, polyfluoroacenes, B, N-substituted polyacenes, and polycyanodienes. The l_{HOMO} and l_{LUMO} values decrease with an increase in molecular size in polyacenes, polyfluoroacenes, B, N-substituted polyacenes, and polycyanodienes. Electron density per atom in the HOMO becomes lower with an increase in molecular size, and the orbital interactions between two adjacent atoms become weaker with an increase in number of atoms in polyacenes, polyfluoroacenes, B, N-substituted polyacenes, and polycyanodienes. Therefore, strengths of the orbital interactions between two adjacent atoms less significantly change with an increase in number of atoms when the monocations and monoanions are distorted along the C–C stretching modes around 1500 cm^{-1} playing an essential role in the electron–phonon interactions. This is the reason why the l_{HOMO} value decreases with an increase in number of atoms. Therefore, in general, we can expect that monocations and monoanions, in which number of carriers per atom is larger, affords larger l_{HOMO} value. The l_{HOMO} and l_{LUMO} values for polyfluoroacenes are larger than those for polyacenes. Larger values of the electron–phonon coupling constants for the C–C stretching modes around 1500 cm^{-1} in the monocations and monoanions of polyfluoroacenes than those in the monocations and monoanions of polyacenes due to the larger displacements of carbon atoms in the C–C stretching modes in polyfluoroacenes than those in polyacenes, are main reason why the l_{HOMO} and l_{LUMO} values for polyfluoroacenes are much larger than those for polyacenes. The l_{HOMO} values for B, N-substituted polyacenes are larger than those for polyacenes. The l_{LUMO} values for B, N-substituted polyacenes are similar to those for polyacenes. The l_{LUMO} values are larger than the l_{HOMO} values in polyacenes, polyfluoroacenes, and B, N-substituted polyacenes. Therefore, electron doping rather than hole doping is more effective way to seek for large electron–phonon coupling constants in polyacenes, polyfluoroacenes, and B, N-substituted polyacenes. Both the l_{HOMO} and l_{LUMO} values for polycyanodienes are larger than those for polyacenes. Therefore, CH–CF, CC–BN, and CC–CN atomic substitutions are effective way to seek for large l_{HOMO} and l_{LUMO} values in polyacenes. The orbital patterns difference between the LUMO localized on carbon atoms located at the edge part of carbon framework in polyacenes and the delocalized LUMO in polycyanodienes due to electronegativity perturbation, is the main reason why the l_{LUMO} values for polycyanodienes are much larger than those for polyacenes. Strong σ orbital interactions between two neighboring carbon atoms and two neighboring carbon and nitrogen atoms and the orbital interactions between two neighboring carbon and hydrogen atoms in the HOMO of polycyanodienes are the main reason why the l_{HOMO} values for σ-conjugated polycyanodienes are much larger than those for π-conjugated polyacenes. The l_{HOMO} values are larger than the l_{LUMO} values in polycyanodienes. Strong σ orbital interactions between two neighboring carbon atoms and two neighboring carbon and nitrogen atoms and the orbital interactions between two neighboring carbon and hydrogen atoms in the HOMO and the weak π orbital interactions between two neighboring carbon atoms and two neighboring carbon and nitrogen atoms in the LUMO of polycyanodienes, are the main reason why the l_{HOMO} values are larger than the l_{LUMO} values in polycyanodienes.

Chapter VII

THE LOGARITHMICALLY AVERAGED PHONON FREQUENCIES

Let us next look into the logarithmically averaged phonon frequencies $\omega_{\ln}$ for the monocations $(\omega_{\ln,\text{HOMO}})$ and monoanions $(\omega_{\ln,\text{LUMO}})$ of polyacenes, polyfluoroacenes, B, N-substituted polyacenes, and polycyanodienes which measure the frequency of the vibrational modes which play an important role in the electron–phonon interactions. The $\omega_{\ln,\text{HOMO}}$ values are defined by

$$\omega_{\ln,\text{HOMO}} = \exp\left\{\sum_m \frac{l_{\text{HOMO}}(\omega_m)\ln\omega_m}{l_{\text{HOMO}}}\right\} \tag{67}$$

and the $\omega_{\ln,\text{LUMO}}$ values are defined by

$$\omega_{\ln,\text{LUMO}} = \exp\left\{\sum_m \frac{l_{\text{LUMO}}(\omega_m)\ln\omega_m}{l_{\text{LUMO}}}\right\} \tag{68}$$

The logarithmically averaged phonon frequencies for the monoanions $(\omega_{\ln,\text{LUMO}})$ and cations of polyacenes, polyfluoroacenes, B, N-substituted polyacenes, and polycyanodienes as a function of molecular weight M_{w} are shown in Figure 12.

A. Polyacenes and Polyfluoroacenes

1. Monocations

The $\omega_{\ln,\text{HOMO}}$ values are estimated to be 1060, 1303, 1341, 1305, and 1270 cm^{-1} for 1fa, 2fa, 3fa, 4fa, and 5fa, respectively, and those are estimated to be 1164, 1521, 1501, 1450, and 1359 cm^{-1} for 1a, 2a, 3a, 4a, and 5a, respectively. Therefore, the $\omega_{\ln,\text{HOMO}}$ values decrease with an increase in molecular size in polyacenes (from 2a to 5a) and polyfluoroacenes (from 2fa to 5fa). This is in qualitatively agreement with a tendency in conventional superconductors; light masses will lead to higher values of $\omega_{\ln}$.

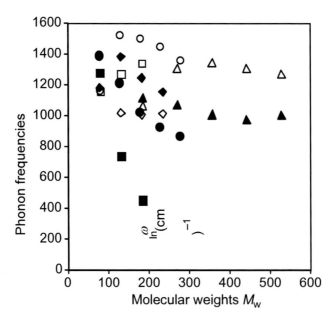

Figure 12. Logarithmically averaged phonon frequencies versus molecular weights in the monocations and monoanions of polyacenes, polyfluoroacenes, B, N-substituted polyacenes, and polycyanodienes. The opened circles, triangles, squares, and diamonds represent the $\omega_{\ln,\text{HOMO}}$ values for polyacenes, polyfluoroacenes, B, N-substituted polyacenes, and polycyanodienes, respectively, and the closed circles, triangles, squares, and diamonds represent the $\omega_{\ln,\text{LUMO}}$ values for polyacenes, polyfluoroacenes, B, N-substituted polyacenes, and polycyanodienes, respectively.

The $\omega_{\ln,\text{HOMO}}$ values for polyfluoroacenes become smaller than those for polyacenes by H–F substitution in polyacenes. This can be understood as follows. The frequencies of the vibronic active modes in polyfluoroacenes are much smaller than those in polyacenes, as expected. Furthermore, the low frequency mode around 500 cm^{-1} as well as the C–C stretching modes around 1500 cm^{-1} can strongly couple to the HOMO in polyfluoroacenes, while only the C–C stretching modes around 1500 cm^{-1} can strongly couple to the HOMO in polyacenes. This is the reason why the $\omega_{\ln,\text{HOMO}}$ values for polyacenes are larger than those for polyfluoroacenes.

2. Monoanions

Let us next look into the logarithmically averaged phonon frequencies $\omega_{\ln}$ for the monoanions of polyfluoroacenes. The $\omega_{\ln,\text{LUMO}}$ values are estimated to be 1112, 1070, 1005, 972, and 1003 cm^{-1} for 1fa, 2fa, 3fa, 4fa, and 5fa, respectively, and those are estimated to be 1390, 1212, 1023, 926, and 869 cm^{-1} for 1a, 2a, 3a, 4a, and 5a, respectively. Therefore, the $\omega_{\ln,\text{LUMO}}$ values for polyacenes and polyfluoroacenes decrease with an increase in molecular size. This is in qualitatively agreement with a tendency in conventional superconductors; light mass will lead to higher values of $\omega_{\ln}$. But it should be noted that the $\omega_{\ln,\text{LUMO}}$ values for polyfluoroacenes decrease with an increase in molecular size less rapidly than those for polyacenes. This can be understood as follows. In the monoanions of polyacenes, the high frequency modes and the low frequency modes, less and more, respectively, strongly couple to the LUMO with an increase in molecular size, and thus, the $\omega_{\ln,\text{LUMO}}$ values significantly decrease with an increase in molecular size. In the monoanions of polyfluoroacenes, on the other hand, the low frequency modes as well as the high frequency modes less strongly couple to the LUMO with an increase in molecular size, and thus, the $\omega_{\ln,\text{LUMO}}$ values slightly decrease with an increase in molecular size. The $\omega_{\ln,\text{LUMO}}$ values for 1fa, 2fa, and 3fa become smaller than those for 1a, 2a, and 3a, respectively, by H–F substitution in polyacenes. However, considering the large M_{w} values difference between 1fa ($M_{\text{w}} = 186$) and 1a ($M_{\text{w}} = 78$), between 2fa ($M_{\text{w}} = 272$) and 2a ($M_{\text{w}} = 128$), and between 3fa ($M_{\text{w}} = 358$) and 3a ($M_{\text{w}} = 178$), the $\omega_{\ln,\text{LUMO}}$ values difference between polyacenes and polyfluoroacenes is not large. This can be understood as follows. Even though the frequencies of all vibronic active modes in polyfluoroacenes are much smaller than those in polyacenes, the displacements of hydrogen and fluorine atoms do not play an essential role in the electron–phonon interactions in the monoanions of polyacenes and polyfluoroacenes, respectively, because the LUMOs are localized on carbon atoms and the electron density on hydrogen and fluorine atoms are very small in polyacenes and polyfluoroacenes, respectively. This is the reason why the $\omega_{\ln,LUMO}$ values difference between polyacenes and polyfluoroacenes is not large even though the M_w values difference between them is large. Furthermore, the $\omega_{\ln,LUMO}$ values for 4fa and 5fa are larger than those for 4a and 5a even though the M_w values for 4fa ($M_w = 444$) and 5fa ($M_w = 530$) are much larger than those for 4a ($M_w = 228$) and 5a ($M_w = 278$), respectively. This is because the low frequency modes as well as the C–C stretching modes around 1500 cm^{-1} strongly couple to the LUMO in 4a and 5a, while the C–C stretching modes around 1500 cm^{-1} much more strongly couple to the LUMO than the low frequency modes in 4fa and 5fa.

We can expect that in the hydrocarbon molecular systems, the $\omega_{\ln}$ values would basically decrease by substituting hydrogen atoms by heavier atoms. This can be understood from the fact that the frequencies of all vibronic active modes in polyacenes downshift by H–F substitution. However, considering that the $\omega_{\ln}$ value for the LUMO rather localized on carbon atoms in 4fa and 5fa becomes larger by H–F substitution, we can expect that the $\omega_{\ln}$ value for a molecular orbital localized on carbon atoms has a possibility to increase by substituting hydrogen atoms by heavier atoms if the orbital patterns of the molecular orbital do not significantly change by such substitution. Therefore, the detailed properties of the vibrational modes and the electronic structures as well as the molecular weights are closely related to the frequencies of the vibrational modes which play an important role in the electron–phonon interactions in the monoanions of polyfluoroacenes.

The $\omega_{\ln,HOMO}$ values are larger than the $\omega_{\ln,LUMO}$ values in polyfluoroacenes. This can be understood as follows. The frequency modes lower than 500 cm^{-1} and the high frequency modes around 1400 cm^{-1} much more strongly couple to the LUMO than to the HOMO in 2fa, 3fa, 4fa, and 5fa, while the frequency modes around 500 cm^{-1} and the frequency modes around 1600 cm^{-1} much more strongly couple to the HOMO than to the LUMO in 2fa, 3fa, 4fa, and 5fa. In particular, the C–C stretching modes around 1600 cm^{-1} the most strongly couple to the HOMO, while the C–C stretching modes around 1400 cm^{-1} the most strongly couple to the

LUMO in 2fa, 3fa, 4fa, and 5fa. This is the reason why the $\omega_{\text{ln,HOMO}}$ values are larger than the $\omega_{\text{ln,LUMO}}$ values in polyfluoroacenes. As described above, the significant phase patterns difference between the HOMO and the LUMO is the main reason why the vibrational modes which play an essential role in the electron–phonon interactions in the monocations are significantly different from those in the monoanions in 2fa, 3fa, 4fa, and 5fa. Therefore, the detailed properties of the vibrational modes and the electronic structures as well as the molecular weights are closely related to the frequencies of the vibrational modes which play an important role in the electron–phonon interactions in charged polyfluoroacenes.

B. B, N-Substituted Polyacenes

The $\omega_{\text{ln,HOMO}}$ ($\omega_{\text{ln,LUMO}}$) values are estimated to be 1154, 1268, and 1337 cm^{-1} (1273, 737, and 449 cm^{-1}) for 1bn, 2bn, and 3bn, respectively. Therefore, the $\omega_{\text{ln,HOMO}}$ and $\omega_{\text{ln,LUMO}}$ values, increase and significantly decrease, respectively, with an increase in molecular size from 1bn to 3bn. As in polyacene-series, the $\omega_{\text{ln,HOMO}}$ values are larger than the $\omega_{\text{ln,LUMO}}$ values in large size of B, N-substituted polyacene-series. And as in 1a, the $\omega_{\text{ln,LUMO}}$ value is slightly larger than the $\omega_{\text{ln,HOMO}}$ value in 1bn.

Let us next compare the calculated results for B, N-substituted polyacene-series with those for polyacene-series. The differences between the $\omega_{\text{ln,HOMO}}$ and $\omega_{\text{ln,LUMO}}$ values become larger with an increase in molecular size in B, N-substituted polyacene-series, while those in polyacene-series hardly change. This can be understood as follows. Let us first look into polyacene-series. [64] The C–C stretching modes around 1500 cm^{-1} and the low frequency modes strongly couple to the LUMO, and less and more, respectively, strongly couple to the LUMO with an increase in molecular size in polyacene-series. On the other hand, the C–C stretching modes around 1500 cm^{-1} strongly couple to the HOMO, and less strongly couple to the HOMO with an increase in molecular size in polyacene-series. Therefore, both the $\omega_{\text{ln,LUMO}}$ and $\omega_{\text{ln,HOMO}}$ values decrease with an increase in molecular size, and the difference between the $\omega_{\text{ln,HOMO}}$ and $\omega_{\text{ln,LUMO}}$ values hardly changes in polyacene-series. Let us next look into B, N-substituted polyacene-series. The B–N stretching modes around 1500 cm^{-1} and the low frequency modes strongly couple to the LUMO, and less and more, respectively, strongly couple to the LUMO with an increase in molecular size in B, N-substituted polyacene-series. On the other hand, the B–N stretching modes around 1500 cm^{-1} and the low frequency modes strongly couple to the HOMO, and the low frequency modes much less strongly couple to the HOMO than the B–N stretching modes around 1500 cm^{-1} with an increase in molecular size in B, N-substituted polyacene-series. This is the reason why the $\omega_{\ln,HOMO}$ and $\omega_{\ln,LUMO}$ values, increase and significantly decrease, respectively, with an increase in molecular size from 1bn to 3bn, and the reason why the differences between the $\omega_{\ln,HOMO}$ and $\omega_{\ln,LUMO}$ values significantly increase with an increase in molecular size in B, N-substituted polyacene-series. Such different properties in the $\omega_{\ln,HOMO}$ and $\omega_{\ln,LUMO}$ values between polyacene-series and B, N-substituted polyacene-series may come from the electronegativity perturbation on polyacene-series.

The $\omega_{\ln,HOMO}$ values for polyacene-series are larger than those for B, N-substituted polyacene-series. This is because only the C–C stretching modes around 1500 cm^{-1} play an essential role in the electron–phonon interactions in polyacene-series, while the low frequency modes as well as the B–N stretching modes around 1500 cm^{-1} play an important role in the electron–phonon interactions in B, N-substituted polyacene-series. The results for the $\omega_{\ln,HOMO}$ values in B, N-substituted polyacene-series are not in agreement with a tendency in solids, as described above. That is, the electronic structures rather than the molecular weights are closely related to the frequencies of the vibrational modes which play an important role in the electron–phonon interactions in the monocations of B, N-substituted polyacene-series. The $\omega_{\ln,HOMO}$ value increases with an increase in molecular size from 1a to 2a, but the $\omega_{\ln,HOMO}$ value slightly decreases with an increase in molecular size from 2a to 6a. On the other hand, the $\omega_{\ln,HOMO}$ values for B, N-substituted polyacene-series increase with an increase in molecular size from 1bn to 3bn. But from analogy with the calculated results for polyacene-series, the $\omega_{\ln,HOMO}$ values may begin to decrease at the finite size of B, N-substituted polyacene-series, according to the general tendency in solids; light mass will lead to higher values of $\omega_{\ln}$.

C. Polycyanodienes

The $\omega_{\ln,HOMO}$ values are estimated to be 1015, 1002, and 1009 cm^{-1} in 2cn, 3cn, and 4cn, respectively. Those are estimated to be 1521, 1501, and 1450 cm^{-1} in 2a, 3a, and 4a, respectively. Therefore, the $\omega_{\ln,HOMO}$ values do not significantly change with an increase in molecular size in polycyanodienes. Furthermore, the $\omega_{\ln,HOMO}$ values for polycyanodienes are smaller than those for polyacenes.

The $\omega_{\ln,LUMO}$ values are estimated to be 1179, 1381, 1242, and 1150 cm^{-1} for 1cn, 2cn, 3cn, and 4cn, respectively, and those are estimated to be 1390, 1212, 1023, and 926 cm^{-1} for 1a, 2a, 3a, and 4a, respectively. Therefore, apart from 1cn, the $\omega_{\ln,LUMO}$ values for polyacenes and polycyanodienes decrease with an increase in molecular weights. This is in qualitative agreement with a tendency; light mass will lead to higher values of $\omega_{\ln}$. But it should be noted that the $\omega_{\ln,LUMO}$ values for polycyanodienes with D_{2h} geometry are larger than those for polyacenes with D_{2h} geometry. As described above, the LUMO of polyacenes is rather localized on carbon atoms located at the edge part of carbon framework in polyacenes, and thus the non-bonding characteristics are significant. Therefore, the orbital interactions between two neighboring carbon atoms are weak. On the other hand, due to electronegativity perturbation, the LUMO of polycyanodienes is delocalized, and the electron density on carbon atoms as well as on nitrogen atoms located at the edge part of CN framework is high, and thus the orbital interactions between two neighboring carbon and nitrogen atoms and between two neighboring carbon atoms in the LUMO are strong. This is the reason why the C–N and C–C stretching modes around 1500 cm^{-1} in the monoanions of polycyanodienes afford larger electron–phonon coupling constants than the C–C stretching modes around 1500 cm^{-1} in the monoanions of polyacenes, and the reason why the $\omega_{\ln,LUMO}$ values for polycyanodienes are larger than those for polyacenes. Therefore, the orbital patterns difference between the delocalized LUMO in polycyanodienes and the LUMO localized on carbon atoms located at the edge part of carbon framework in polyacenes is the main reason why the l_{LUMO} and $\omega_{\ln,LUMO}$ for polycyanodienes are larger than those for polyacenes.

D. Summary

In this chapter, we investigated the logarithmically averaged phonon frequencies which measure the frequencies of the vibronic active modes playing an essential role in the electron–phonon interactions in polyacenes, polyfluoroacenes, B, N-substituted polyacenes, and polycyanodienes. The $\omega_{\mathrm{ln,HOMO}}$ and $\omega_{\mathrm{ln,LUMO}}$ values decrease with an increase in molecular size in polyacenes and polyfluoroacenes. This is in qualitative agreement with a tendency in conventional superconductors; light mass will lead to higher values of ω_{ln}. The $\omega_{\mathrm{ln,HOMO}}$ values for polyfluoroacenes become smaller than those for polyacenes by H–F substitutions. The $\omega_{\mathrm{ln,LUMO}}$ values for polyfluoroacenes decrease with an increase in molecular size less rapidly than those for polyacenes. The $\omega_{\mathrm{ln,LUMO}}$ values for 1fa, 2fa, and 3fa become smaller than those for 1a, 2a, and 3a, by H–F substitutions. However, considering the large M_{w} values difference between 1fa and 1a, 2a and 2fa, and 3a and 3fa, the $\omega_{\mathrm{ln,LUMO}}$ values difference between polyacenes and polyfluoroacenes is not large. Furthermore, the $\omega_{\mathrm{ln,LUMO}}$ values for 4fa and 5fa are larger than those for 4a and 5a even though the M_{w} values for 4fa and 5fa are larger than those for 4a and 5a, respectively. We can expect that in the hydrocarbon molecular systems, the ω_{ln} values would basically decrease by substituting hydrogen atoms by heavier atoms. This can be understood from the fact that the frequencies of all vibronic active modes in polyacenes downshift by H–F substitution. However, considering that the ω_{ln} value for the LUMO rather localized on carbon atoms in 4fa and 5fa becomes larger by H–F substitution, we can expect that the ω_{ln} value for a molecular orbital localized on carbon atoms has a possibility to increase by substituting hydrogen atoms by heavier atoms if the phase patterns of the molecular orbital do not significantly change by such atomic substitution. Therefore, the detailed properties of the vibrational modes and the electronic structures as well as the molecular weights are closely related to the frequencies of the vibronic active modes playing an important role in the electron–phonon interactions in the monoanions of polyfluoroacenes. The $\omega_{\mathrm{ln,HOMO}}$ values are larger than the $\omega_{\mathrm{ln,LUMO}}$ values in polyfluoroacenes. The significant phase patterns difference between the HOMO and LUMO is the main reason why the frequencies of vibronic active modes playing an essential role in the electron–phonon interactions in the monocations are significantly different from those in the monoanions in 2fa, 3fa, 4fa, and 5fa. Therefore, the detailed properties of the vibrational modes and the electronic states as well as the molecular weights are closely related to the frequencies of the vibronic active modes playing an important role in the electron–phonon interactions in charged polyfluoroacenes. The $\omega_{\ln,HOMO}$ and $\omega_{\ln,LUMO}$ values, increase and significantly decrease, respectively, with an increase in molecular size in B, N-substituted polyacenes. As in polyacenes, the $\omega_{\ln,HOMO}$ values are larger than the $\omega_{\ln,LUMO}$ values in large size of B, N-substituted polyacenes. As in 1a, the $\omega_{\ln,HOMO}$ value is slightly larger than the $\omega_{\ln,LUMO}$ value in 1bn. The differences between the $\omega_{\ln,HOMO}$ and $\omega_{\ln,LUMO}$ values become larger with an increase in molecular size in B, N-substituted polyacenes, while those in polyacenes hardly change. Such different properties in the $\omega_{\ln,HOMO}$ and $\omega_{\ln,LUMO}$ values between polyacenes and B, N-substituted polyacenes may come from the electronegativity perturbation on polyacenes. The $\omega_{\ln,HOMO}$ values for polyacenes are larger than those for B, N-substituted polyacenes. The results for the $\omega_{\ln,HOMO}$ values in B, N-substituted polyacenes are not in agreement with a tendency in solids, as described above. That is, the electronic structures rather than the molecular weights are closely related to the frequencies of the vibrational modes playing an important role in the electron–phonon interactions in the monocations of B, N-substituted polyacenes. The $\omega_{\ln,HOMO}$ values do not significantly change with an increase in molecular size in polycyanodienes. Furthermore, the $\omega_{\ln,HOMO}$ values for polycyanodienes are smaller than those for polyacenes. Apart from 1cn, the $\omega_{\ln,LUMO}$ values for polyacenes and polycyanodienes decrease with an increase in molecular weights. The $\omega_{\ln,LUMO}$ values for D_{2h} symmetric polycyanodienes are larger than those for D_{2h} symmetric polyacenes. The phase patterns difference between the delocalized LUMO in polycyanodienes and LUMO localized on carbon atoms located at the edge part of carbon framework in polyacenes is the main reason why the $\omega_{\ln,LUMO}$ values for polycyanodienes are larger than those for polyacenes.

Chapter VIII

CONCLUDING REMARKS

In this book, we discussed the electron–phonon interactions in the charged molecular systems such as polyacenes, polyfluoroacenes, B, N-substituted polyacenes, and polycyanodienes. We estimated the electron–phonon coupling constants and the frequencies of the vibronic active modes playing an essential role in the electron–phonon interactions. These physical values are essential to discuss several physical phenomena such as intramolecular electrical conductivity, intermolecular charge transfer, attractive electron–electron interactions and Bose–Einstien condensation, and superconductivity, which will be discussed in detail in the next review article. Motivated by the possible inverse isotope effects in Pd-H, Pd-D, and Pd-T superconductivity, and organic superconductivity observed by Saito et al., we discussed how the H–F substitution are closely related to the essential characteristics of the electron–phonon interactions in these molecules by comparing the calculated results for charged polyacenes with those for charged polyfluoroacenes, since fluorine atoms are much heavier than D and T atoms, and the phase patterns of the frontier orbitals such as the HOMO and LUMO are not expected to be significantly changed. Furthermore, we discuss how CC–BN and CC–CN substitutions are closely related to the essential characteristics of the electron–phonon interactions in these molecules by comparing the calculated results for charged polyacenes with those for charged B, N-substituted polyacenes and polycyanodienes, respectively. These physical values are essential to discuss the several physical phenomena such as intramolecular electrical conductivity, intermolecular charge transfer, attractive electron–electron interactions and Bose–Einstien condensation, and superconductivity, which will be discussed in detail in the next review article.

In this review article, we investigated the electron–phonon interactions in the monocations and monoanions of polyacenes, polyfluoroacenes, B, N-substituted polyacenes, and polycyanodienes. The C–C stretching modes around 1500 cm^{-1} strongly couple to the HOMO, and the lowest frequency modes and the C–C stretching modes around 1500 cm^{-1} strongly couple to the LUMO in polyacenes. The C–C stretching modes around 1500 cm^{-1} strongly couple to the HOMO and LUMO in polyfluoroacenes. The B–N stretching modes around 1500 cm^{-1} strongly couple to the HOMO and LUMO in B, N-substituted polyacenes. The C–C and C–N stretching modes around 1500 cm^{-1} strongly couple to the HOMO and LUMO in polycyanodienes.

We estimated the total electron–phonon coupling constants in the monocations and monoanions of polyacenes, polyfluoroacenes, B, N-substituted polyacenes, and polycyanodienes. The l_{HOMO} and l_{LUMO} values decrease with an increase in molecular size in polyacenes, polyfluoroacenes, B, N-substituted polyacenes, and polycyanodienes. In general, we can expect that monocations and monoanions, in which number of carriers per atom is larger, affords larger l_{HOMO} value. The l_{HOMO} and l_{LUMO} values for polyfluoroacenes are larger than those for polyacenes. The l_{HOMO} values for B, N-substituted polyacenes are larger than those for polyacenes. The l_{LUMO} values for B, N-substituted polyacenes are similar to those for polyacenes. The l_{LUMO} values are larger than the l_{HOMO} values in polyacenes, polyfluoroacenes, and B, N-substituted polyacenes. Therefore, electron doping rather than hole doping is more effective way to seek for large electron–phonon coupling constants in polyacenes, polyfluoroacenes, and B, N-substituted polyacenes. Both the l_{HOMO} and l_{LUMO} values for polycyanodienes are larger than those for polyacenes. Therefore, CH–CF, CC–BN, and CC–CN atomic substitutions are effective way to seek for large l_{HOMO} and l_{LUMO} values in polyacenes. The orbital patterns difference between the LUMO localized on carbon atoms located at the edge part of carbon framework in polyacenes and the delocalized LUMO in polycyanodienes due to electronegativity perturbation, is the main reason why the l_{LUMO} values for polycyanodienes are much larger than those for polyacenes. Strong σ orbital interactions between two neighboring carbon atoms and two neighboring carbon and nitrogen atoms and the orbital interactions between two neighboring carbon and hydrogen atoms in the HOMO of polycyanodienes are the main reason why the l_{HOMO} values for σ-conjugated polycyanodienes are much larger than those for π-conjugated polyacenes. The l_{HOMO} values are larger than the l_{LUMO} values in polycyanodienes. Strong σ

orbital interactions between two neighboring carbon atoms and two neighboring carbon and nitrogen atoms and the orbital interactions between two neighboring carbon and hydrogen atoms in the HOMO and the weak π orbital interactions between two neighboring carbon atoms and two neighboring carbon and nitrogen atoms in the LUMO of polycyanodienes, are the main reason why the l_{HOMO} values are larger than the l_{LUMO} values in polycyanodienes.

We also investigated the logarithmically averaged phonon frequencies which measure the frequencies of the vibronic active modes playing an essential role in the electron–phonon interactions in polyacenes, polyfluoroacenes, B, N-substituted polyacenes, and polycyanodienes. The $\omega_{ln,HOMO}$ and $\omega_{ln,LUMO}$ values decrease with an increase in molecular size in polyacenes and polyfluoroacenes. This is in qualitative agreement with a tendency in conventional superconductors; light mass will lead to higher values of ω_{ln}. The $\omega_{ln,HOMO}$ values for polyfluoroacenes become smaller than those for polyacenes by H–F substitutions. The $\omega_{ln,LUMO}$ values for polyfluoroacenes decrease with an increase in molecular size less rapidly than those for polyacenes. The $\omega_{ln,LUMO}$ values for 1fa, 2fa, and 3fa become smaller than those for 1a, 2a, and 3a, by H–F substitutions. However, considering the large M_w values difference between 1fa and 1a, 2a and 2fa, and 3a and 3fa, the $\omega_{ln,LUMO}$ values difference between polyacenes and polyfluoroacenes is not large. Furthermore, the $\omega_{ln,LUMO}$ values for 4fa and 5fa are larger than those for 4a and 5a even though the M_w values for 4fa and 5fa are larger than those for 4a and 5a, respectively. We can expect that in the hydrocarbon molecular systems, the ω_{ln} values would basically decrease by substituting hydrogen atoms by heavier atoms. This can be understood from the fact that the frequencies of all vibronic active modes in polyacenes downshift by H–F substitution. However, considering that the ω_{ln} value for the LUMO rather localized on carbon atoms in 4fa and 5fa becomes larger by H–F substitution, we can expect that the ω_{ln} value for a molecular orbital localized on carbon atoms has a possibility to increase by substituting hydrogen atoms by heavier atoms if the phase patterns of the molecular orbital do not significantly change by such atomic substitution. Therefore, the detailed properties of the vibrational modes and the electronic structures as well as the molecular weights are closely related to the frequencies of the vibronic active modes playing an important role in the electron–phonon interactions in the monoanions of polyfluoroacenes. The $\omega_{ln,HOMO}$ values are larger than the $\omega_{ln,LUMO}$ values in polyfluoroacenes. The significant phase patterns difference between the HOMO and LUMO is the main reason why the frequencies of vibronic active modes playing an essential role in the electron–phonon interactions in the monocations are significantly different from those in the monoanions in 2fa, 3fa, 4fa, and 5fa. Therefore, the detailed properties of the vibrational modes and the electronic states as well as the molecular weights are closely related to the frequencies of the vibronic active modes playing an important role in the electron–phonon interactions in charged polyfluoroacenes. The $\omega_{\text{ln,HOMO}}$ and $\omega_{\text{ln,LUMO}}$ values, increase and significantly decrease, respectively, with an increase in molecular size in B, N-substituted polyacenes. As in polyacenes, the $\omega_{\text{ln,HOMO}}$ values are larger than the $\omega_{\text{ln,LUMO}}$ values in large size of B, N-substituted polyacenes. As in 1a, the $\omega_{\text{ln,HOMO}}$ value is slightly larger than the $\omega_{\text{ln,LUMO}}$ value in 1bn. The differences between the $\omega_{\text{ln,HOMO}}$ and $\omega_{\text{ln,LUMO}}$ values become larger with an increase in molecular size in B, N-substituted polyacenes, while those in polyacenes hardly change. Such different properties in the $\omega_{\text{ln,HOMO}}$ and $\omega_{\text{ln,LUMO}}$ values between polyacenes and B, N-substituted polyacenes may come from the electronegativity perturbation on polyacenes. The $\omega_{\text{ln,HOMO}}$ values for polyacenes are larger than those for B, N-substituted polyacenes. The results for the $\omega_{\text{ln,HOMO}}$ values in B, N-substituted polyacenes are not in agreement with a tendency in solids, as described above. That is, the electronic structures rather than the molecular weights are closely related to the frequencies of the vibrational modes playing an important role in the electron–phonon interactions in the monocations of B, N-substituted polyacenes. The $\omega_{\text{ln,HOMO}}$ values do not significantly change with an increase in molecular size in polycyanodienes. Furthermore, the $\omega_{\text{ln,HOMO}}$ values for polycyanodienes are smaller than those for polyacenes. Apart from 1cn, the $\omega_{\text{ln,LUMO}}$ values for polyacenes and polycyanodienes decrease with an increase in molecular weights. The $\omega_{\text{ln,LUMO}}$ values for D_{2h} symmetric polycyanodienes are larger than those for D_{2h} symmetric polyacenes. The phase patterns difference between the delocalized LUMO in polycyanodienes and LUMO localized on carbon atoms located at the edge part of carbon framework in polyacenes is the main reason why the $\omega_{\text{ln,LUMO}}$ values for polycyanodienes are larger than those for polyacenes.

In summary, CH–CF, CC–BN, and CC–CN atomic substitutions are effective way to seek for larger l_{HOMO} values, and CH–CF and CC–CN atomic substitutions are effective way to seek for larger l_{LUMO} values in polyacenes.

ACKNOWLEDGMENT

This study was performed under the Project of Academic Frontier Center at Nagasaki Institute of Applied Science. This work is partly supported by a Grant-in-Aid for Scientific Research from the Japan Society for the Promotion of Science (JSPS-16560618, JSPS-17350094).

REFERENCES

[1] (a) I. B. Bersuker, *The Jahn–Teller Effect and Vibronic Interactions in Modern Chemistry* (Plenum, New York, 1984); (b) I. B. Bersuker and V. Z. Polinger, *Vibronic Interactions in Molecules and Crystals* (Springer, Berlin, 1989); (c) I. B. Bersuker, *Chem. Rev.* **101**, 1067 (2001); (d) I. B. Bersuker, *The Jahn–Teller Effect* (Cambridge University Press., Cambridge, 2006).

[2] G. Grimvall, *The Electron-Phonon Interaction in Metals* (North-Holland, Amsterdam, 1981).

[3] G. Fischer, *Vibronic Coupling: The Interaction between the Electronic and Nuclear Motions* (Academic, London, 1984).

[4] (a) N. O. Lipari, C. B. Duke, and L. Pietronero, *J. Chem. Phys.* **65**, 1165 (1976); (b) P. Pulay, G. Fogarasi, and J. E. Boggs, *J. Chem. Phys.* **74**, 3999 (1981).

[5] Text books, (a) C. Kittel, *Quantum Theory of Solids* (Wiley, New York, 1963); (b) J. M. Ziman, *Principles of the Theory of Solids* (Cambridge University, Cambridge, 1972); (c) H. Ibach and H. Lüth, *Solid-State Physics* (Springer, Berlin, 1995).

[6] (a) J. R. Schrieffer, *Theory of Superconductivity* (Addison-Wesley, Massachusetts, 1964); (b) P. G. de Gennes, *Superconductivity of Metals and Alloys* (Benjamin, New York, 1966).

[7] W. A. Little, *Phys. Rev.* **134**, A1416 (1964).

[8] (a) D. Jérome, A. Mazaud, M. Ribault, and K. Bechgaad, *J. Phys.* (France) *Lett.* **41**, L95 (1980); (b) M. Ribault, G. Benedek, D. Jérome, and K. Bechgaad, ibid. **41**, L397 (1980).

[9] Reviews and books: (a) D. Jérome and H. J. Schulz, *Adv. Phys.* **31**, 299 (1982); (b) T. Ishiguro and K. Yamaji, *Organic Superconductors* (Springer, Berlin, 1990); (c) J. M. Williams, J. R. Ferraro, R. J. Thorn, K. D. Karlson,

U. Geiser, H. H. Wang, A. M. Kini, and M. -H. Whangbo, *Organic Superconductors* (Prentice Hall, New Jersey, 1992).

[10] K. Oshima, H. Urayama, H. Yamochi, and G. Saito, *J. Phys. Soc. Jpn.* **57**, 730 (1988).

[11] E. Demiralp, S. Dasgupta, and W. A. Goddard III, *J. Am. Chem. Soc.* **117**, 8154 (1995).

[12] (a) A. F. Hebard, M. J. Rosseinsky, R. C. Haddon, D. W. Murphy, S. H. Glarum, T. T. M. Palstra, A. P. Ramirez, and A. R. Kortan, *Nature* **350**, 600 (1991); (b) M. J. Rosseinsky, A. P. Ramirez, S. H. Glarum, D. W. Murphy, R. C. Haddon, A. F. Hebard, T. T. M. Palstra, A. R. Kortan, S. M. Zahurak, and A. V. Makhija, *Phys. Rev. Lett.* **66**, 2830 (1991).

[13] K. Tanigaki, T. W. Ebbesen, S. Saito, J. Mizuki, J. S. Tsai, Y. Kubo, and S. Kuroshima, *Nature* **352**, 222 (1991).

[14] T. T. M. Palstra, O. Zhou, Y. Iwasa, P. E. Sulewski, R. M. Fleming, and B. R. Zegarski, *Solid State Commun.* **93**, 327 (1995).

[15] (a) C. M. Varma, J. Zaanen, and K. Raghavachari, *Science* **254**, 989 (1991); (b) M. Lannoo, G. A. Baraff, M. Schlüter, and D. Tomanek, *Phys. Rev. B* **44**, 12106 (1991); (c) Y. Asai and Y. Kawaguchi, *Phys. Rev. B* **46**, 1265 (1992); (d) J. C. R. Faulhaber, D. Y. K. Ko, and P. R. Briddon, *Phys. Rev. B* **48**, 661 (1993); (e) V. P. Antropov, O. Gunnarsson, and A. I. Lichtenstein, *Phys. Rev. B* **48**, 7651 (1993); (f) A. Auerbach, N. Manini, and E. Tosatti, *Phys. Rev. B* **49**, 12998 (1994); (g) N. Manini, E. Tosatti, and A. Auerbach, *Phys. Rev. B* **49**, 13008 (1994); (h) O. Gunnarsson, *Phys. Rev. B* **51**, 3493 (1995); (i) O. Gunnarsson, H. Handschuh, P. S. Bechthold, B. Kessler, G. Ganteför, and W. Eberhardt, *Phys. Rev. Lett.* **74**, 1875 (1995); (j) J. L. Dunn and C. A. Bates, *Phys. Rev. B* **52**, 5996 (1995); (k) O. Gunnarsson, *Rev. Mod. Phys.* **69**, 575 (1997); (l) A. Devos and M. Lannoo, *Phys. Rev. B* 58, 8236 (1998); (m) O. Gunnarsson, *Nature* **408**, 528 (2000).

[16] E. A. Silinch and V. Capek, *Organic Molecular Crystals* (AIP, New York, 1994).

[17] W. Warta and N. Karl, *Phys. Rev. B* **32**, 1172 (1985).

[18] W. Warta, R. Stehle, and N. Karl, *Appl. Phys. A* **36**, 163 (1985).

[19] S. Kivelson and O. L. Chapman, *Phys. Rev. B* **28**, 7236 (1983).

[20] A. Mishima and M. Kimura, *Synth. Met.* **11**, 75 (1985).

[21] T. Skoskiewicz, *Phys. Status Solidi* **A11**, K123 (1972).

[22] B. Stritzker and W. Buckel, *Z. Physik* **257**, 1 (1972).

[23] W -H. Li, J. W. Lynn, H. B. Stanley, T. J. Udovic, R. N. Shelton, and P. Klavins, *Phys. Rev. B* **257**, 4119 (1989).

[24] J. E. Schirber, J. M. Mintz, and W. Wall, *Solid State Commun.* **52**, 837 (1984).

[25] (a) T. Kato and T. Yamabe, *J. Chem. Phys.* **115**, 8592 (2001); (b) T. Kato and T. Yamabe, *Recent Research Developments in Quantum Chemistry* (Transworld Research Network, Kerala, 2004); (c) *Recent Research Developments in Physical Chemistry* (Transworld Research Network, Kerala, 2004).

[26] V. Coropceanu, M. Malagoli, D. A. da Silva Filho, N. E. Gruhn, T. G. Bill, and J. L. Brédas, *Phys. Rev. Lett.* **89**, 275503 (2002).

[27] (a) T. Kato and T. Yamabe, J. Chem. Phys. 120, 7659 (2004); (b) T. Kato and T. Yamabe, *J. Chem. Phys.* **119**, 11318 (2003).

[28] (a) T. Kato and T. Yamabe, Chem. Phys. 315, 97 (2005); (b) T. Kato and T. Yamabe, *J. Chem. Phys.* **118**, 3300 (2003).

[29] (a) T. Kato and T. Yamabe, J. Chem. Phys. 123, 094701 (2005); (b) T. Kato and T. Yamabe, *J. Phys. Chem. A* **108**, 11223 (2004).

[30] (a) T. A. Albright, J. K. Burdett, and M. -H. Whangbo, *Orbital Interactions in Chemistry* (Wiley, New York, 1985); (b) J. K. Burdett, *Chemical Bonding in Solids* (Oxford University, Oxford, 1995).

[31] E. M. Conwell, *Phys. Rev. B* **22**, 1761 (1980).

[32] K. Tanaka, Y. Huang, and T. Yamabe, *Phys. Rev. B* **51**, 12715 (1995).

[33] C. Christides, D. A. Neumann, K. Prassides, J. R. D. Copley, J. J. Rush, M. J. Rosseinsky, D. W. Murphy, and R. C. Haddon, *Phys. Rev. B* **46**, 12088 (1992).

[34] (a) A. D. Becke, *Phys. Rev. A* **38**, 3098 (1988); (b) *J. Chem. Phys.* **98**, 5648 (1993).

[35] C. Lee, W. Yang, and R. G. Parr, *Phys. Rev. B* **37**, 785 (1988).

[36] (a) R. Ditchfield, W. J. Hehre, and J. A. Pople, J. Chem. Phys. 54, 724 (1971); (b) P. C. Hariharan and J. A. Pople, *Theor. Chim. Acta* **28**, 213 (1973).

[37] M. J. Frisch, G. W. Trucks, H. B. Schlegel, P. M. W. Gill, B. G. Johnson, M. A. Robb, J. R. Cheeseman, T. A. Keith, G. A. Petersson, J. A. Montgomery, K. Raghavachari, M. A. Al-Laham, V. G. Zakrzewski, J. V. Ortiz, J. B. Foresman, J. Cioslowski, B. B. Stefanov, A. Nanayakkara, M. Challacombe, C. Y. Peng, P. Y. Ayala, W. Chen, M. W. Wong, J. L. Andres, E. S. Replogle, R. Gomperts, R. L. Martin, D. J. Fox, J. S. Binkley, D. J. Defrees, J. Baker, J. J. P. Stewart, M. Head-Gordon, C. Gonzalez, and J. A. Pople, *Gaussian* **98**, (Gaussian Inc., Pittsburgh, Pennsylvania, 1998).

[38] C. R. Patrick and G. S. Prosser, *Nature* **187**, 1021 (1960).

[39]   W. A. Duncan, J. P. Sheridan and F. L. Swinton, *Trans. Faraday Soc.* **62**, 1090 (1966).

[40]   J. R. Goates, J. B. Ott and J. Reeder, *J. Chem. Thermodyn.* **5**, 135 (1973).

[41]   J. S. Brennen, N. M. D. Brown and F. L. Swinton, *J. Chem. Soc. Faraday Trans.* **1** 70, 1965 (1974).

[42]   J. B. Ott, J. R. Goates and D. L. Carbon, *J. Chem. Thermodyn.* **8**, 505 (1976).

[43]   T. Dahl, *Acta Chem. Scand.* **25**, 1031 (1971).

[44]   T. Dahl, *Acta Chem. Scand.* **26**, 1569 (1972).

[45]   J. M. Steed, T. A. Dixon, and W. Klemperer, *J. Chem. Phys.* **70**, 4940 (1979).

[46]   J. H. Williams, *Mol. Phys.* **73**, 99 (1991).

[47]   J. H. Williams, *Chem. Phys.* **172**, 171 (1993).

[48]   M. Neelakandan, D. Pant, E. L. Quitevis, *Chem. Phys. Lett.* **265**, 283 (1997).

[49]   M. Neelakandan, D. Pant, E. L. Quitevis, *J. Phys. Chem. A* **101**, 2936 (1997).

[50]   C. G. Gray and K. E. Gubbins, *Theory of Molecular Fluids* (Clarendon Press, New York, 1984; Vol. 1, p 587).

[51]   J. R. Riddick and W. B. Bunger, *Organic Solvents* (Wiley-Inter-science, New York, 1970).

[52]   M. R. Battaglia, A. D. Buckingham, and J. H. Williams, *Chem. Phys. Lett.* **78**, 421 (1981).

[53]   (a) N. Del Campo, M. Besnard, and J. Yarwood, *Chem. Phys.* 184, 225 (1994); (b) M. Besnard, Y. Danten, and T. Tassaing, In *Collision- and Interaction-Induced Spectroscopy* G. C. Tabisz and M. N. Neuman Eds. (Kluwer, Dordrecht, 1995; p 201-213).

[54]   (a) F. Williams, M. B. Yim, and D. E. Wood, *J. Am. Chem. Soc.* **95**, 6475 (1973); (b) M. B. Yim and D. E. Wood, *J. Am. Chem. Soc.* **98**, 2053 (1976); (c) M. Shiotani and F. Williams, *J. Am. Chem. Soc.* 98, 4006 (1976); (d) M. C. R. Symons, R. C. Selby, J.G. Smith, and S. W. Bratt, Chem. *Phys. Lett.* **48**, 100 (1977); (e) J. T. Wang and F. Williams, *Chem. Phys. Lett.* **71**, 471 (1980); (f) O. A. Anisimov, V. M. Grigoriants, and Yu. N. Molin, *Chem. Phys. Lett.* **74**, 15 (1980); (g) L. N. Shchegoleva, I. I. Bilkis, and P. V. Schastnev, *Chem. Phys.* **82**, 343 (1983); (h) W. E. Wentworth, T. Limero, and E. C. M. Chen, *J. Phys. Chem.* **91**, 241 (1987); (i) P. V. Schastnev and L. N. Shchegoleva, *Molecular Distortions in Ionic and Excited States* (CRC Press, Boca Raton, FL, 1995, p. 79); (j) K. Hiraoka, S. Mizume, and S. Yamabe, *J. Phys. Chem.* **94**, 3689 (1990); (k) L. C. T. Shoute and J. P.

Mittal, *J. Phys. Chem.* **97**, 379 (1993); (l) A. Hasegawa, M. Shiotani, and Y. Hama, *J. Phys. Chem.* **98**, 1834 (1994); (m) L. N. Shchegoeva, I. I. Beregovaya, and P. V. Schastnev, *Chem. Phys. Lett.* **312**, 325 (1999).

[55] (a) W. D. Hobey and A. D. Mclachlan, *J. Chem. Phys.* **33**, 1695 (1960); (b) L. N. Shchegoeva and P. V. Schastnev, *Chem. Phys. Lett.* **130**, 115 (1989); (c) J. J. Nash and R. R. Squires, *J. Am. Chem. Soc.* **118**, 11872 (1996); (d) V. V. Konovalov, S. S. Laev, I. V. Beregovaya, L. N. Shchegoleva, V. D. Shteingarts, Yu, D. Tsvetkov, and I. I. Bilkis, *J. Phys. Chem. A* **104**, 152 (2000).

[56] I. B. Bersuker, *Chem. Rev.* **101**, 1067 (2001).

[57] (a) T. A. Miller, *Annu. Rev. Phys. Chem.* **33**, 257 (1982); (b) T. J. Sears, T. A. Miller, and V. E. Bondybey, *J. Chem. Phys.* **74**, 3240 (1981).

[58] E. J. M. Hamilton, S. E. Dolan, C. M. Mann, H. O. Colijn, C. A. McDonald, and S. G. Shore, *Science* **260**, 659 (1993).

[59] J. -Y. Yi and J. Bernholc, *Phys. Rev. B* **47**, 1708 (1993).

[60] A. Rubio, J. L. Corkill, and M. L. Cohen, *Phys. Rev. B* **49**, 5081 (1994).

[61] Y. Miyamoto, A. Rubio, M. L. Cohen, and S. G. Louie, *Phys. Rev. B* **50**, 4976 (1994).

[62] Y. Miyamoto, A. Rubio, S. G. Louie, and M. L. Cohen, *Phys. Rev. B* **50**, 18360 (1994).

[63] X. Blase, A. Rubio, S. G. Louie, and M. L. Cohen, *Europhys. Lett.* **28**, 335 (1994).

[64] X. Blase, J. -C. Charlier, A. De Vita, and R. Car, *Appl. Phys. Lett.* **70**, 197 (1997).

[65] B. -C. Wang, M. -H. Tsai, and Y.-M. Chou, *Synth. Met.* **86**, 2379 (1997).

[66] M. Menon and D. Srivastava, *Chem. Phys. Lett.* **307**, 407 (1999).

[67] D. L. Carroll, P. Redlich, P. M. Ajayan, S. Curran, S. Roth, and M. Rühle, *Carbon* **36**, 753 (1998).

[68] X. Blase, J. -C. Charlier, A. De Vita, and R. Car, *Appl. Phys. A* **68**, 293 (1999).

[69] P. W. Fowler, K. M. Rogers, G. Seifert, M. Terrones, and H. Terrones, *Chem. Phys. Lett.* **299**, 359 (1999).

[70] V. V. Pokropivny, V. V. Skorokhod, G. S. Oleinik, A. V. Kurdyumov, T. S. Bartnitskaya, A. V. Pokropivny, A. G. Sisonyuk, and D. M. J. Scheichenko, *Solid State Chem.* **154**, 214 (2000).

[71] T. Hirano, T. Oku, and K. Suganuma, *Diamond and Related Materials* 9, 625 (2000).

[72] S. Erkoc, *J. Mol. Struc. (THEOCHEM)* **542**, 89 (2001).

[73] J. Kongsted, A. Osted, L. Jensen, P. Åstrand, and K. V. Mikkelsen, *J. Phys. Chem. B* **105**, 10243 (2001).

[74] H. Steinberg and R. J. Brotherton, *Organoboron Chemistry* (Wiley, New York, 1966, Vol. 2).

[75] A. Meller, Top. *Cum. Chem.* **15**, 146 (1970).

[76] W. Maringgele, *The Chemistry of Inorganic Homo- and Heterocycles* (I. Haiduc, Ed.; Academic Press, London, 1987).

[77] D. F. Gaines and J. Borlin, *Boron Hydride Chemistry* (E. L. Muet- teries, Ed.; Academic Press, New York, 1975, p 241).

[78] K. Niedenzu and J. W. Dawson, *Boron-Nitrogen Compounds* (Academic Press, New York, 1965).

[79] R. T. Paine and C. K. Narula, *Chem. Rev.* **90**, 73 (1990).

[80] A. Stock and E. Pohland, *Chem. Rer.* **59**, 2215 (1926).

[81] (a) S. G. Shore and R. W. Parry, *J. Am. Chem. Soc.* **77**, 6084 (1955); (b) S. G. Shore and R.W. Parry, *J. Am. Chem. Soc.* **80**, 8 (1958); (c) S. G. Shore and K. W. Bddeker, *Inorg. Chem.* **3**, 914 (1964).

[82] A. W. Laubengayer, P. C. Jr. Moews, and R. F. Porter, *J. Am. Chem. Soc.* **83**, 1337 (1961).

[83] Y. Matsunaga, *Bussei Kagaku* (In Japanese) (Syokabo, Tokyo, 1981).

[84] J. L. Brédas, B. Thémans, and J. M. André, *J. Chem. Phys.* **78**, 6137 (1983).

INDEX

A

amplitude, 18, 28, 29
atomic orbitals, 21, 23, 26, 38
atoms, viii, 4, 21, 23, 26, 28, 31, 33, 34, 35,
    36, 37, 38, 39, 42, 43, 44, 45, 46, 50, 52,
    53, 56, 57, 58, 59, 60, 66, 69, 70, 73, 74, 75

B

BCS theory, 2
bending, 27, 31, 39
benzene, 18, 19, 25
bonding, 21, 28, 31, 32, 34, 35, 37, 38, 39, 42,
    43, 44, 45, 56, 59, 69
bonds, 16, 26, 39

C

carbon, viii, 21, 23, 25, 26, 28, 31, 32, 34, 35,
    36, 37, 38, 43, 44, 45, 47, 50, 52, 53, 56,
    57, 58, 59, 60, 66, 69, 70, 74, 75
carbon atoms, viii, 21, 23, 28, 31, 32, 34, 35,
    36, 37, 38, 45, 50, 52, 53, 56, 57, 59, 60,
    66, 69, 70, 74, 75
carbon nanotubes, 25
cation, 23
cell, 17
C-N, 25

compensation, 37, 38
condensation, 4, 73
conductivity, 4, 73
consensus, 1
control, 1
couples, 28, 31, 33, 34, 35, 36, 37, 38, 40, 42,
    43, 45, 46, 52
coupling, vii, 1, 2, 4, 5, 6, 7, 8, 9, 10, 11, 12,
    15, 16, 17, 18, 19, 27, 28, 29, 30, 31, 32,
    33, 34, 36, 37, 39, 40, 41, 42, 43, 44, 46,
    49, 50, 51, 53, 55, 57, 58, 60, 69, 73, 74
coupling constants, vii, 2, 4, 6, 7, 9, 10, 11,
    12, 15, 16, 27, 29, 30, 31, 32, 33, 34, 37,
    39, 40, 41, 42, 43, 44, 46, 49, 50, 51, 53,
    55, 57, 58, 60, 69, 73, 74
crystals, 1, 2, 15

D

deformation, 1
degenerate, 1, 7, 9, 10, 18, 25
density, 16, 21, 23, 26, 38, 40, 50, 53, 56, 57,
    59, 60, 66, 69
derivatives, 7, 27
DFT, 21, 27
diamonds, 51, 64
distortions, 57
doping, 60, 74

E

electrical conductivity, 1, 4, 73
electron density, 23, 38, 56, 59
electronic structure, viii, 52, 66, 67, 68, 70, 75
electronic systems, 1
electron–phonon interactions, vii, 1, 2, 4, 37, 38, 42, 50, 60, 67, 68, 70, 73, 75
electrons, 1, 5
energy, 5, 6, 8, 9, 10, 12, 19, 21, 23, 25, 26, 27, 28, 31, 33, 34, 35, 37, 38, 40, 42, 43, 45, 56, 57
equilibrium, 7, 16, 17, 19, 27
exciton, 1
expectations, 2
expression, 2, 16
extrapolation, 53

F

Fermi level, 1, 16, 17
fluorine, 4, 23, 31, 32, 34, 35, 36, 37, 38, 52, 66, 73
fluorine atoms, 4, 23, 31, 32, 34, 35, 36, 37, 38, 52, 66, 73
France, 79

H

Hamiltonian, 5
hybrid, 21, 27
hydrides, 2
hydrocarbons, 21
hydrogen, viii, 1, 2, 26, 36, 40, 43, 44, 52, 57, 58, 61, 66, 70, 74, 75
hydrogen atoms, viii, 26, 36, 40, 43, 44, 52, 57, 58, 61, 66, 70, 74, 75
hypothesis, 2

I

instability, 1
interaction, 1, 8, 21, 42

interactions, vii, 1, 2, 4, 5, 10, 15, 16, 21, 26, 28, 31, 32, 34, 35, 37, 38, 39, 42, 43, 44, 45, 46, 47, 50, 53, 56, 57, 58, 59, 60, 63, 66, 67, 68, 69, 70, 73, 74, 75
ionization, 2
isotope, 1, 2, 4, 73

J

Japan, 77

M

mass, 31, 35, 36, 40, 46, 65, 68, 69, 70, 75
matrix, 5, 7, 8, 10, 12, 16, 17, 18, 19
mode, 1, 7, 8, 9, 10, 11, 12, 15, 16, 17, 18, 19, 27, 28, 31, 32, 34, 35, 36, 37, 38, 39, 40, 41, 43, 44, 45, 46, 50, 54, 65
molecular structure, 1
molecular weight, viii, 63, 64, 66, 67, 68, 69, 70, 75
molecules, vii, 1, 4, 16, 17, 21, 23, 73

N

naphthalene, 2, 25
neglect, 17
New York, iv
nitrogen, 25, 26, 45, 47, 57, 58, 59, 61, 69, 74
nitrogen compounds, 25
nuclei, 5

P

pairing, 15
palladium, 2
phonons, 1
photoelectron spectroscopy, 2
physical properties, 23
physics, 1
polarizability, 23
polyacenes, vii, 2, 4, 5, 6, 7, 11, 12, 13, 16, 17, 20, 21, 22, 23, 24, 25, 26, 27, 29, 30,

36, 37, 40, 41, 42, 47, 49, 50, 51, 52, 54,
55, 56, 57, 58, 59, 60, 63, 64, 65, 66, 69,
70, 73, 74, 75, 76
polycyanodienes, vii, 4, 5, 6, 7, 13, 16, 17, 20,
22, 24, 26, 44, 46, 47, 49, 51, 57, 58, 59,
60, 63, 64, 69, 70, 73, 74, 75
prediction, 1
pressure, 2
program, 21
pyrolysis, 25

R

range, 27
resolution, 2

S

scattering, 15
series, 39, 42, 43, 53, 55, 56, 58, 59, 67, 68
shape, 23
sign, 23
spectroscopy, 1
spin, 16, 18
stabilization, 36, 38
stretching, vii, 27, 28, 31, 34, 35, 36, 37, 39,
40, 41, 42, 43, 44, 46, 47, 50, 52, 53, 54,
55, 57, 60, 65, 66, 67, 68, 69, 74
substitution, viii, 2, 4, 23, 25, 37, 52, 65, 66,
70, 73, 75
superconductivity, 1, 2, 4, 15, 73

superconductor, 1
symmetry, 7, 11, 23, 25, 26
synthesis, 1
systems, vii, 1, 2, 4, 15, 23, 66, 70, 73, 75

T

theory, 1, 2, 16, 21, 26, 31
thermolysis, 25
time, 23
total energy, 7
transformations, 19
transition, 1, 2
transition temperature, 1, 2
transport, 2

V

values, vii, 4, 15, 49, 50, 51, 52, 53, 55, 56,
57, 58, 59, 60, 63, 64, 65, 66, 67, 68, 69,
70, 73, 74, 75, 76
variation, 21, 23, 26
vibration, 8, 11

W

wave number, 27
wave vector, 16, 17
workers, 25